BUSINESS FINANCE

WAS NEVER THIS EASY

HARVEY GOLDSTEIN

ISBN: 978-1-63945-319-1 (Paperback)

978-1-63945-320-7 (Ebook)

The views expressed in this book are solely those of the author and do not necessarily reflect the views of the publisher, and the publisher hereby disclaims any responsibility for them.

Writers' Branding
1800-608-6550
www.writersbranding.com
orders@writersbranding.com

*Dedication
To my wife*

Table of Contents

Business Finance Was Never This Easy is written in concise and plain, non-technical language. It takes the reader through the accounting basics, but then proceeds to move into the analyses, measuring sticks, investment analysis (including own company financial valuation), and product cost. It also discusses the three levels of cash flow; the key to every business, and financial planning.

But beyond the discussion of these mechanics/techniques, the book also discusses policy decisions and the potential impacts of these policies. It presents and discusses the mechanics, raises and discusses questions regarding financial policies, and provides other thought provoking material.

The information presented within Business Finance follows basic financial principles, which provides financial information to enhance the reader's understanding of the "business end of business."

The presentations are made for the sake of clarity to individuals not heavily schooled in finance; but who want and need the understanding of it. Further, the presentations are prepared, and reflect the methods and systems developed through my thirty plus years of experience. Any similarity to other systems is a result of a combination of basic financials and comparable experience.

Harvey Goldstein

ACKNOWLEDGEMENT

There have been many people that have provided input into this book. The technical aspects of this book came from my head... years of experience. Several people have contributed to review and enhancement. To them may I say thank you for your time, expertise and genuine interest in this project... Bill Franzen and Catheen Faulx; and to my wife Enid who contributed to ensuring that it was understandable.

SECTION ONE
The Basics

CHAPTER ONE —
A FEW PRIVATE MOMENTS WITH HARVEY

Over the course of my extensive career in various financial positions, owners and managers often came into my office requesting an explanation of certain financial situations or the financial impact of a decision or both. In this regard, it had been recommended to them to spend "a few private moments with Harvey."

When I first started my career, I understood that my function was to prepare financial statements, perform analyses, develop budgets, and execute financial planning, among other duties. I thought this was the most critical part of my job - to know and apply the mechanics. I soon learned it was equally important to explain what was happening in an easily understandable fashion. I know now it is in this capacity, the capacity to explain what is going on that a financial manager truly earns their money.

To most people, finance is a dry subject and to some extent, I agree. To me, however, counseling that ensures individuals at all levels of an organization understand the financial ramification of the business is fun. Admittedly, it always helps when the manager has some understanding of finance. But, it is not always necessary. Even if the individual is numerically challenged, as I have been told, financial functions are more easily grasped after spending "a few private moments with Harvey." As an instructor at the University of Phoenix, my experiences teaching both undergraduate and graduate finance courses to my students (many who are not majoring in finance) reaffirms my belief that finance can be a lot more fun when it is understood.

We are a society who loves scores and score cards; be it your golf game, bowling strikes, runs in a baseball game, counting your daily calories, or your exercise routines. As a capitalistic society, we want to know how we are doing. Numbers are the common denominator in almost every discipline. Numbers in the business world tell us if the company is being accepted by the public, if it is efficiently managed, and most important, if it is making money. In other words, this is the company's scorecard.

So, will this book or "a few private moments with Harvey" make you a financial genius? No! Do you want to become a financial genius? Probably not. But, if you would like to have a good understanding of the financial aspects of your business or, if you would like to impress your co-workers and boss with your finance savvy, then this is the book for you.

This book is presented in a user-friendly fashion. For example, each section will answer the following three questions:

1. What do I need to know and why do I need to know it?
2. How do I use the information to better manage a business?
3. What must I do to understand the financial side of a business?

Within the book are useful formulas and examples that are presented for your use in an easy to understand fashion. These are some of the items that an owner, manager, or student really needs to know in order to analyze and maximize the value of a business.

When you examine the <u>Table of Contents</u>, you will readily see the flow of the book. The first section addresses the basics or answers the question, "Oh, is that how it works?" The accounting is there for you to simply understand how it operates. No need for panic. It is nothing more than that.

Section Two, <u>Understanding and Planning</u>, is not as hard as it may appear at first glance. This section sets the groundwork to help you manage and monitor a business. You will become undaunted by the financial operating requirements, financial reviews and tests, as well as by financial reporting and forecasting; and by what may very well be the most critical issue, cash flow. You will be able to plan for the future and to view different strategies to potentially maximize your performance. This is a major key to the success of any business. You know...the fun stuff.

In Section Three, <u>Analysis</u>, the reader is introduced to the exciting challenges of investment analysis and returns, profitability measures, production costs, critical management controls, and valuing your company. Scary sounding? Not really. With the knowledge you will have acquired from each previous section, you will be comfortable with each new concept and be able to understand it more easily.

The bottom line is that the goals of "Finance for the (Busy and Disinterested) Businessman" are to enable the reader to:

1. Understand the basics of financial management;
2. Financially plan and test different assumptions; and
3. Manage and understand a business better using certain tried and true financial techniques.

Periodically, I will allude to situations I have personally encountered that illustrate and enhance a point. While these "war stories" may or may not directly relate to your current situation, they will point to pitfalls to be avoided. Take the time to read them, they will be well worth your time.

Before I begin, I would like to emphasize two items. I once visited a company where financial management consisted of two checking accounts, one for operations and the other as a reserve. When funds flowed into the reserve account, all was fine. However that was not always the case, you know the rest. Unfortunately, this was a contributing factor to the business' demise. Note two items:

1. Putting cash into a business is not revenue. While there are other sources of income, the company's basic business operation produces the revenue and is the basis for its existence.
2. Revenue is generated from sales from regular business operations.

Now you're ready, so let's go.

CHAPTER TWO —
CHART OF ACCOUNTS

Introduction to Financial Statements

A system of organizing your finance operations.

An account is a grouping of like revenues or expenditures initiating from business operations. The source could be your (1) checkbook (funds received or disbursed), (2) various documents for items received and promises to pay, or (3) shipments made and collections to be received.

The Chart of Accounts is the organization of a business' accounts. It groups the accounts into: (1) asset accounts, (2) liability accounts, (3) equity accounts, (4) revenue accounts, (5) cost accounts, and (6) expense accounts. Briefly: asset accounts reflect what you own; liability accounts detail what you owe; equity accounts reflect the net worth of the company; revenue accounts detail what you sell (either a good or service); cost accounts detail the price necessary to generate the revenue, and expense accounts reflect how and what it takes to manage and support the business. The groups are setup in sufficient detail to provide a story of how you performed and how much you are worth.

Most Charts of Accounts use account numbers to group and organize the accounts similar to the listing and grouping below. Note that the initial organization is into two principal financial reports: the Balance Sheet and Income Statement. The definition of each type of account placed into each report or statement will be stated in Chapter Four, Financial Statements.

<u>Balance Sheet Accounts</u>

1000 Assets
2000 Liabilities
3000 Equity accounts

<u>Income Statement</u>

4000 Revenues
5000 Cost of sales
6000 Selling expenses
7000 General & Administrative Expenses
8000 Open
9000 Other Income and Expenses

Within these categories, sub-groupings and specific accounts are set up to collect and classify the various activities.

Assets

For example within the Assets on the **Balance Sheet,** there would be current assets, long term assets, and other assets. Current assets are those whose life is less than one year. Within the current assets are such actual accounts as cash, accounts receivables, inventories, prepaid expenses; such as, business insurance fully or partially paid in advance and not yet "earned" by the insurance company. This will be discussed later. Long term assets accounts (in excess of one year in economic life) are accounts for plant, property and equipment; such as, buildings, equipment, vehicles, computer equipment, and even computer software. Also in that grouping would be the accumulated depreciation accounts taken in each category. This will be further discussed in Chapter Five. This is in order for one to easily determine, on the books, the net value of each asset group, after the depreciation. This is called the Net Book Value. "Other Assets" would be Goodwill, Patents, and other such non-tangible assets that cannot be placed into any other asset sub-group. Goodwill is the value of assets that are in excess of the estimated market value at time of purchase.

Liabilities

For **Liabilities**, again there are current liabilities (those which must be paid within one year of their incurrence) and long-term liabilities. The current liabilities are usually: trade accounts payable, notes payable, wages payable, taxes payable; short term loans (less than one year to maturity), the current portion of long term loans or capital leases. The "current" portion reflects those amounts that must be paid within one-year of the statement date. Long-term liabilities include all loans (including private, banks, and publicly traded securities and bonds) in excess of one year to maturity. However, that portion of each loan that is not due for payment for at least one year from the date of the statement is considered "long term." This is usually the situation with installment loans. Also in long term liabilities are long term loans to individuals. An investigation of the loan document will reveal how the accounting should be handled.

Equity

The **Equity** accounts are in the Equity/Net Worth Section of the Balance Sheet. There are accounts for: common stock (the amount received for the issuance of ownership stock), additional paid-in capital (the amount received in excess of the stated "par" values of each share), retained earnings (amount of profit not distributed to the shareholders from earnings in prior periods), and current year net earnings. Dividends are "paid" from the retained earnings account. However, a separate account under the retained earnings account may be desired, just to easily follow such distributions.

For the **Income Statement**, accounts include sales, cost of goods sold, or cost of sales. Various other accounts are created detailing all expenses; such as salaries, wages, payroll taxes, auto expense, rent, advertising, utilities, office expense, small tools, and supplies. And, as many other accounts as is necessary to track and understand the operations of the business.

There may be more than one sales account. This is done to track the sales of different products or services; all within the "4000" grouping of accounts.

The inclusion and grouping of all accounts depend on the company's operations, how the company thinks (its culture), and how your industry generally sets up its chart of accounts. The industry's normal Chart of Accounts has usually evolved from experience in doing its business and reporting to the industry's management and owners. For example, the cost of goods sold accounts (or cost of sales accounts) could include direct materials (even a separation of major items), direct labor, associated payroll taxes, and direct equipment rental—provided that these items are regularly utilized in the production of what is sold.

The **"expenses"** may be organized into sub-groups to provide additional clarity. For example, instead of including advertising, sales commissions, sales travel, promotions, trade shows, etc. into the General and Administrative Expenses (G & A) category, they may be sufficiently large and under one manager to make another sub-group called "Sales and Marketing." While the G & A expenses may or may not separate sales and marketing expenses, it would also include such accounts as salaries (for management), telephone, rent, various insurances, travel, entertainment, etc.

In summary, a Chart of Accounts is set up to separate and report what the company owns, what the company owes, and what the company is worth—which is the difference between assets and liabilities. Also, the chart is used to track how the company is operating; its sales, costs and expenses, and resulting earnings or loss. Various separations and groupings are made of similar accounts for the purpose of ease of understanding. A numbering system is set up to further organize, identify, and group these accounts in a meaningful manner.

In Chapter Four, Financial Statements, a more detailed Chart of Accounts will be presented.

CHAPTER THREE —
ACCOUNTING: THE VERY BASICS

Basic accounting: The goal in this brief section is to familiarize the reader with the actual flow of the data through an accounting system. It will enable the reader to picture the actual accounting entries that end up in any account. The accounts are grouped (per Chapter Two) and generate financial statements. This chapter is not designed to instruct the reader in accounting by any stretch of the imagination.

An account is a grouping of like revenues or expenditures initiating from business operations. The source could be (1) your checkbook (funds received or disbursed) or (2) from various documents for (a) items received, (b) promises to pay, (c) shipments made, or (d) collections to be received.

Accounts are set up or created in as much detail as is desired by management and required to satisfy legal regulations. The accounting system is referred to as a "double-entry" system. This means that each time financial activity is recorded into the accounting system, it has, at least, two items equaling each other. All "entries" consist of a left side, called the "debit," and a right side, called a "credit." By definition, a receipt of cash is a debit or left side entry into the books. Since there are two sides to every entry, the offset is a right side or credit entry. From this definition, one should be able to determine for each entry what will be done.

Let's start with several examples (we will first describe the transaction, then show the actual accounting entry) in which the transaction is first, then the actual accounting entry will be shown.

Harvey Goldstein

Please note, good accounting practice requires that each entry be followed by a reason for the entry. It also requires supporting documentation; such as, a cash register receipt, a packing slip signed as received by the customer, an own company receiving report for the receipt of anything coming into the business and an invoice, etc.

Example # 1: A sale was made for $100.00 on credit. Two weeks later, the invoice is paid.

Date	Description	Debit	Credit
2XXX			
Feb 12	Accounts receivable	100.00	
	Sales		100.00
	To record the sale		
Feb 26	Cash	100.00	
	Accounts Receivable		100.00

We will repeat these entries in the example below, making reference to the same dates to provide continuity and flow into the trial balance.

This message or technique will occur in every instance that an activity is recorded into the financial records of the company.

Example # 2: Our company purchases inventory for cash. The actual activity consists of receiving the inventory and drawing a check for payment. Accounting will debit (left side) the Inventory Account to add to the inventory by the amount purchased. (This increases the inventory value). Then, since it was paid for with a check, it will credit your cash balance by the amount of the check. Again, the numbers must match exactly. So the complete transaction would debit the amount of purchase, and credit the amount of the check (cash). Now both are in balance. The purchase amount is $125.00. See below.

Date	Description	Debit	Credit
2XXX			
Feb 16	Inventory	125.00	
	Cash		125.00
	To record the purchase of inventory		

Other transaction occurring before and after the ones cited above are detailed below.			
Jan 1	Cash	1,000.00	
	Common Stock		1,000.00
To record investment into the XYZ Co. by Mr. "A." Stock certificate #1001			
Jan 1	Machinery	500.00	
	Accounts Payable		500.00
To record the purchase of Widget machine, tag #101 (editors note: to be paid in the future)			
Jan 15	Office Supplies	50.00	
	Cash		50.00
To purchase office supplies for cash.			
Feb 1	Telephone	45.00	
	Accounts Payable		45.00
To record telephone expense; set up for payment			
Feb 1	Lease Expense (forklift)	75.00	
	Accounts Payable		75.00
To record the lease of a forklift			
Feb 2	Inventory	200.00	
	Cash		200.00
To record purchase of Widget holders			
Feb 12	Accounts receivable	100.00	
	Sales		100.00
To record the sale			
Feb 12	Cost of Goods Sold	65.00	
	Inventory		65.00

	To record the cost of inventory sold with order #101 & to relieve (reduce) inventory			
Feb 26	Cash		100.00	
	Accounts receivable			100.00
	To record the receipt of payment for invoice #1			
Feb 28	Salesman's commission		10.00	
	Cash			10.00
	To record the payment of commission for sale #1			
Feb 28	Depreciation expense		6.00	
	Accumulated Depreciation			6.00
	To record the depreciation expense for one month of equipment			

This last entry is an accounting entry to provide for the use of the equipment (depreciation expense). It also reduces the value of the asset accordingly. It will be further discussed in Chapter Five on **Depreciation and Taxes**.

The February 26th entry reflects the collection of cash for the February 12, 2XXX sale on credit (Accounts Receivable [AIR]). Thus, the former $100.00 A/R balance (debit) is offset as the entry places the $100.00 into the cash account and credits A/ R. Also note, the cash should be confirmed or supported by a bank deposit receipt for the $100.00.

These activities always occur within the accounting system. There are entries that are much more complicated where each side of the entry has a number of line items. However, they all must total to exactly the same amount on each side.

Placing the activity in the wrong account causes mistakes, even though the entry is in balance. This will be discovered in reviewing the account's activity and in the audit process.

Each activity or entry is posted to their respective accounts in the ledger. All the above entries are recorded below. Thus, we are creating the ledger. As it contains all the accounts, it is called the General Ledger. There are special supporting ledgers, generally, for accounts payable, accounts receivable, and fixed assets. These support the balances within the general ledger for those accounts. The sub ledgers are created when there is too much activity to be detailed in the general ledger account.

<table>
<tr><td colspan="7" align="center">XYZ Company
General Ledger
As of February 28, 2XXX</td></tr>
<tr><td colspan="7">(Note: within each account are "posted" each transaction from the above journal entries. This provides traceability to the original transaction. Also note the account numbers, per Chapter Three</td></tr>
<tr><td colspan="2" align="center">Account number; name</td><td colspan="2" align="center">Activities</td><td colspan="2" align="center">Balance</td></tr>
<tr><td>Date</td><td>Item</td><td>Debit</td><td>Credit</td><td>Debit</td><td>Credit</td></tr>
<tr><td colspan="2">1001; Cash</td><td></td><td></td><td></td><td></td></tr>
<tr><td>1</td><td>Jan sale of common stock</td><td>1,000.00</td><td></td><td></td><td></td></tr>
<tr><td>15</td><td>Jan office supplies</td><td></td><td>50.00</td><td></td><td></td></tr>
<tr><td>2</td><td>Feb purchase inventory</td><td></td><td>200.00</td><td></td><td></td></tr>
<tr><td>16</td><td>Feb purchase inventory</td><td></td><td>125.00</td><td></td><td></td></tr>
<tr><td>25</td><td>Feb accounts receivable</td><td>100.00</td><td></td><td></td><td></td></tr>
<tr><td>28</td><td>Feb commission payment</td><td></td><td>10.00</td><td>715.00</td><td></td></tr>
<tr><td colspan="2">1100; Inventory</td><td></td><td></td><td></td><td></td></tr>
<tr><td>2</td><td>Feb cash</td><td>200.00</td><td></td><td></td><td></td></tr>
<tr><td>12</td><td>Feb cost of goods sold</td><td></td><td>65.00</td><td></td><td></td></tr>
<tr><td>16</td><td>Feb cash</td><td>125.00</td><td></td><td>260.00</td><td></td></tr>
<tr><td colspan="2">1500; Machinery & Equipment</td><td></td><td></td><td></td><td></td></tr>
<tr><td>1</td><td>Jan accounts payable</td><td>500.00</td><td></td><td>500.00</td><td></td></tr>
<tr><td colspan="2">1510 Accumulated depreciation</td><td></td><td></td><td></td><td></td></tr>
<tr><td>28</td><td>Feb depreciation expense</td><td></td><td>6.00</td><td></td><td>(6.00)</td></tr>
<tr><td colspan="2">2010; Accounts Payable</td><td></td><td></td><td></td><td></td></tr>
<tr><td>1</td><td>Jan purchase machinery</td><td></td><td>500.00</td><td></td><td></td></tr>
<tr><td>1</td><td>Feb telephone expense</td><td></td><td>45.00</td><td></td><td></td></tr>
<tr><td>1</td><td>Feb lease expense/forklift</td><td></td><td>75.00</td><td></td><td>(620.00)</td></tr>
<tr><td colspan="2">3010; Common Stock</td><td></td><td></td><td></td><td></td></tr>
<tr><td>1</td><td>Jan</td><td></td><td>1,000.00</td><td></td><td>(1,000.00)</td></tr>
<tr><td colspan="2">4010; Sales</td><td></td><td></td><td></td><td></td></tr>
<tr><td>12</td><td>Feb accounts receivable #1</td><td></td><td>100.00</td><td></td><td>(100.00)</td></tr>
<tr><td colspan="2">5010; Cost of Goods sold</td><td></td><td></td><td></td><td></td></tr>
<tr><td>12</td><td>Feb inventory</td><td>65.00</td><td></td><td>65.00</td><td></td></tr>
<tr><td colspan="2">6050; Depreciation expense</td><td></td><td></td><td></td><td></td></tr>
<tr><td>28</td><td>Feb accumulated depreciation</td><td>6.00</td><td></td><td>6.00</td><td></td></tr>
<tr><td colspan="2">6070; Lease expense</td><td></td><td></td><td></td><td></td></tr>
<tr><td>1</td><td>Feb accounts payable</td><td>75.00</td><td></td><td>75.00</td><td></td></tr>
<tr><td colspan="2">6080; Office supplies</td><td></td><td></td><td></td><td></td></tr>
<tr><td>15</td><td>Jan cash</td><td>10.00</td><td></td><td>10.00</td><td></td></tr>
<tr><td colspan="2">6090; Salesman commission</td><td></td><td></td><td></td><td></td></tr>
<tr><td></td><td>cash</td><td>10.00</td><td></td><td>10.00</td><td></td></tr>
<tr><td colspan="2">6100; Telephone expense</td><td></td><td></td><td></td><td></td></tr>
<tr><td>1</td><td>Feb accounts payable</td><td>45.00</td><td></td><td>45.00</td><td></td></tr>
</table>

Each of the account balances becomes part of a Trial Balance. The Trial Balance is a listing of the balances of every account that has any activity or a balance; an accuracy check to ensure that the total of the accounts are in balance. Each account balance could be either a debit or a credit, and sums up to total for each side. They are in balance when the totals for each side match—to the penny. This does not state that the activities within each account are correct; but that the total of the account balances are equal; the total of the debit balances match the total of the credit account balances.

<table>
<tr><td colspan="4" align="center">XYZ Company
Trial Balance
As of February 28, 2XXX</td></tr>
<tr><th>Number</th><th>Name</th><th>DR</th><th>CR</th></tr>
<tr><td>1001</td><td>Cash</td><td>$715.00</td><td></td></tr>
<tr><td>1050</td><td>Accounts receivable</td><td>0.00</td><td></td></tr>
<tr><td>1100</td><td>Inventory</td><td>260.00</td><td></td></tr>
<tr><td>1500</td><td>Machinery & equipment</td><td>500.00</td><td></td></tr>
<tr><td>1510</td><td>Accumulated depreciation</td><td></td><td>$6.00</td></tr>
<tr><td>2010</td><td>Accounts payable</td><td></td><td>620.00</td></tr>
<tr><td>3010</td><td>Common Stock</td><td></td><td>1,000.00</td></tr>
<tr><td>4010</td><td>Sales</td><td></td><td>100.00</td></tr>
<tr><td>5010</td><td>Cost of goods sold</td><td>65.00</td><td></td></tr>
<tr><td>6050</td><td>Depreciation expense</td><td>6.00</td><td></td></tr>
<tr><td>6070</td><td>Lease expense</td><td>75.00</td><td></td></tr>
<tr><td>6080</td><td>Office supplies</td><td>50.00</td><td></td></tr>
<tr><td>6090</td><td>Salesman's commission</td><td>10.00</td><td></td></tr>
<tr><td>6100</td><td>Telephone expense</td><td>45.00</td><td></td></tr>
<tr><td></td><td>Totals</td><td>$1,726.00</td><td>$1,726.00</td></tr>
</table>

Cash Accounting Vs. Accrual Accounting

According to U.S. tax law, a company has the privilege of choosing whether it prefers to account for its business activity on a cash basis or an accrual basis. This option exists for up to $5 million in annual sales. Above this level, the accrual basis is required.

The Cash basis is just that. Nothing is entered into the financial records regarding earnings unless cash is exchanged. Thus, the sale would not be recorded within the records until the actual collection is made. A detailed record is kept of the accounts and amounts (accounts receivable ledger/aging) owed to the company and accounts that are owed by the company (accounts payable ledger). The Income Statement does not reflect it for earnings purposes- until the cash portion is transacted. All the transactions are kept as detailed as in any other system. However, reiterating, the recording of a sale as a sale doesn't take place until payment is received. The same goes for inventory (the cost of sales). It is not recorded until payment is made for the materials received. If a company buys a ream of paper on credit, it is not recorded as an expense until the company draws the check to pay the vendor. If a company buys widgets for sale but doesn't pay for them in the same period that they are sold (payment is collected), the sale would have no cost charged against it. However, the reverse is also true! If no sale is made in a time period (no actual payment is collected), but disbursements for raw materials are made, then the income statement for that period would reflect no sales, but a large cost of sale; reflecting the payment. In the long run, generally, the costs are matched with the sales.

Summarizing, cash transactions are recorded only when actual cash is exchanged. However, detailed records are kept as in any system of accounting. Thus, under this method, it is possible for the cost of sales to be in excess of sales. This could occur if a large shipment of products is received and shipped to a customer within a month when "sales" (collections) are low. The opposite is also very possible, if in the following month (for example) no (or very little) materials are paid for with strong sales and collections occurring.

The Accrual method includes all transactions for earnings purposes, even if no actual cash is exchanged. The goal is twofold: 1) to match the cost with the sales and 2) to record the obligation as incurred. The transactions examples above were made on this (accrual) basis.

If the company ships goods to a customer, the customer is then obligated to pay our company and the sale is placed on the books. The expectation is that payment will be collected per the terms of the sale. Conversely, when a company buys a ream of paper, the company incurs the expense on its books as well as the obligation to pay. Thus, it will set up the obligation in their official accounts payable records or ledger and show the ream of paper as an expense of doing business when the goods are received.

We have come to the end of the mechanics and flow of the bookkeeping and financial activity recording. In summary, there are two means of accounting for small businesses under $5 million in sales: cash basis and accrual basis. Each time an activity has a financial impact or should be recorded (for tracking or for any reason), an entry is made into the financial records. It is made via a double-sided entry into the books. The two sides ensure against missing any data and balancing all activities.

CHAPTER FOUR —
FINANCIAL STATEMENTS

Our goal in this chapter is to combine all the various accounts into reports to advise the reader of: (1) what the company is worth and why, (2) how to advise, and explain the company's financial performance over a specific period of time, and (3) to identify what is generating cash and where it is being spent.

Definitions: Expenditures, a disbursement of money.

> *Expense. an accounting item wherein expenditure is made as part of business operations, for which the value received will not last beyond one year and is not directly associated with the production of product for sale. It is a reduction of earnings.*

> *Cost. generally, those expenditures made to produce the product that is sold. This includes the cost of the materials taken from inventory, packaged, and ready for shipment on the loading dock to a customer. It is also a charge against earnings.*

When the Chart of Accounts is in place, the accounts are then combined into financial statements. The chart is never really finished as accounts are added or eliminated as the business and information requirements change.

The prior chapters identified the general account categories, and the two basic financial reports. An account by its use will reflect how much it is worth as of a point in time or it will reflect how much business activity has occurred within it, over a period of time. There are two basic financial statements; a Balance Sheet and an Income

Statement. The Balance Sheet details what a company is worth at a point in time. An Income Statement details a company's financial performance over a period of time.

Different accounts are included within each statement. For example, the amount that customers owe the company is an active value or balance as of a point in time. Thus, it is a Balance Sheet account. However, spending on advertising accumulates over a defined period; it does not show a tangible balance (what it is worth) as of the moment, but the activity over a period of time. Thus, it is an Income Statement account; in this case, an expense.

In summary, a *Balance Sheet* account is an account in which there are actual values. These values represent how much the company is owed, how much cash a company has, the value of its inventory, or how much it owes, amongst others. An *Income Statement* account reflects the history of each account over a period of time, but each cannot be transformed into anything of actual value. Its value is that it advises the activity of the account defined over a period of time.

All accounts sum to the revenue generated less the costs and expenses incurred. If the revenue accounts exceed the cost and expense accounts, then, in summary, a profit is produced. That profit adds to the wealth of the company. Repeating: it adds to the wealth of the company. Or, if the Income Statement shows a loss, then it reduces the wealth of the company.

When the earnings statement shows a profit or loss, it is "transferred" to an account on the Balance Sheet (Net Worth section), called current year earnings, which becomes part of the retained earnings accounts. Thus, it affects the worth of the company.

Thus, the Balance Sheet is the sum of all accounts and accounting for a business. In incorporates the results of all Profit and Loss Statements activities, and all other activities of the company. From a pure accounting viewpoint, it ensures that all items and accounts "tie-in" to one another, via individual accounts and through the formula, Assets = Liabilities and Net Worth (just as it appears on the statement).

Balance Sheet

Expanding, the **Balance Sheet** discloses the value of the company at a particular point in time. It outlines what the company **owns** (as in cash, inventory, or property and equipment) and what **it is owed** (accounts receivable, by its customers or from an individual or entity). These are detailed in the ASSETS section of the Balance Sheet statement. It also details what **it owes** (via accounts payable to vendors, employees, individuals, or financial institutions). These are detailed in the LIABILITIES section of the statement. The difference between the assets balance and the liabilities balance is what the company is worth or the net worth of the company.

The Net Worth of the company (or the Stockholders Equity section of the statement) has its own accounts. The accounts are common stock and sometimes preferred stock; additional paid-in capital; retained earnings (prior years' earnings or losses), and current year's earnings. Dividends are another item that may have its own account or it may be incorporated into the retained earnings (reducing as the dividends are paid from or charged to this account). Via the accounting system, each account has activity within it. (See the initial accounting in the prior chapters regarding the receiving of cash via sale of common stock.) Thus, the accounts making up the Net Worth section of the Balance Sheet will equal the difference between the total assets and total liabilities. This is also a factor in determining the value of the company, if it were for sale. But it is only one factor.

The formula that makes up the Balance Sheet statement is: Assets = Liabilities + Net Worth. By definition, the balance sheet must balance to the penny.

The balance sheet could, generally, be structured as follows:

The XYZ Company
Balance Sheet
As of December 31, 2:xxx

Assets
Current Assets
Cash
Accounts Receivable
Inventory
Other current assets

Total current assets
Plant, property and equipment
Less: accumulated depreciation
Net Plant, property and equipment
Loan from Officer : due {beyond one year)

Other Assets
Patents
Goodwill
Total Other Assets
Total Assets

Liabilities
Current Liabilities
Accounts Payable
Accrued payables
Other payables
Short-term loan/line of credit

Long-Term Liabilities
Bank term loan
Bonds
Loan from relative

Total Liabilities

Net Worth/Stockholder Equity
Preferred Stock, at par
Common Stock, at par
Additional paid-in capital
Retained earnings
Current earnings
Total Net Worth

Total Liabilities and Net Worth

Long-Term assets

Total long-term assets

Total current liabilities

Total long-term liabilities

Income Statement

The **Income Statement** is the reflection of how the business performed over a period of time; a month, quarter, or a whole year. The statement below generally reflects a manufacturing or distribution business. Financial statements while strictly adhering to GAAP (Generally Accepted Accounting Principals) may differ for each industry. For example, banking would be significantly different from the income statement shown below. A restaurant would more closely follow the one below as it incorporates a cost of sales similar to a cost of goods sold, followed by general business expenses. In general, an income statement is defined as follows: Revenues - costs - expenses = profit (or loss). See the example below:

Different Industries could not only have "different looking" statements; but also, different relationships between the accounts. These differences can also show up in the balance sheet statement as well.

It should also be noted that even within the same type of business, within the same industry, the Income Statement for each could probably have differing accounts and even a differing Chart of Accounts. This may be due to the "culture" of the business and the information requirements of the management of each business. Below is a "typical" Income Statement.

The XYZ Company
Income Statement
For the year ended December 31, 2xxx

Gross Sales
Less: return, discounts & allowances
Net Sales

Cost of Sales
Direct materials
Direct labor
Factory overhead
Total Cost of Sales

Gross Profit

General and Administrative Expenses
Advertising
Auto expense
Bank fees
Business taxes
Commissions
Depreciation
Office supplies
Rent
Administrative salaries
Payroll taxes
Total General and Administrative Expenses

Operating Earnings

Other Income and Deductions
Interest Income
Interest expense
Bad debts
Total Other Income & Deductions

Earnings Before Income Taxes (EBIT)
Provision for Incomes Taxes

Net Earnings

Thus, via the Balance Sheet and the Income Statement, we will know how the company is performing and why, and the company's net worth.

There is a third statement that is just as critical: **The Cash-Flow Statement**. This may be the most critical statement as it deals with cash—and cash runs a business. The Cash-Flow Statement is a derivation (recombination) of the Income Statement and the Balance Sheet. This statement can take a variety of different formats. However, it generally follows an organization of detailing cash flow from operating activities, investing activities, and financing activities.

Operating activities are the bottom line net earnings on the Income Statement (after provision for taxes) and plus (adding back) the non-cash expenses, such as, depreciation and amortization contained and charged within the Income Statement. It also reflects the changes in certain current assets that would affect the company's cash position, such as accounts receivable, inventory and accounts payable.

Investing activities are those expenditures made for investment by the business. It could be for machinery and equipment, computer equipment, new buildings, rolling stock or short-term or long-term investment of excess funds.

Financing activities are the acquisition of loans to maintain operations or to purchase the equipment acquired in the investment section. It details the securement of loans and their repayments, as well as the distribution of dividends and the inflow of cash from the sale of new equity securities.

This will be further discussed in Chapter Twelve, Cash Flow.

4A.-BASICS REVIEW

We have discussed the very basics of accounting (bookkeeping), and the basics of financial reports. We have created entries into the books, created the general ledger from the entries, and generated a trial balance. Let's actually create a company, we will again go through startup and operations, create the actual entries, and produce and understand the financial statements.

Scenario

You have a wonderful idea to produce and distribute Widgets in two colors. (Widgets are actual equipment found on sailing ships in the mid-19th century. And if there are any sailing ships still operating, they would be found on them too.) We'll call the company The Widget Company, with you as the only stockholder. And to protect yourself, you have chosen it to be an "S" corporation. (Ownership will be discussed in Chapter Six).

The Widget Company will prosper by distributing raw widgets and widgets in two colors. You believe it is the colors of the widgets that are missing from the market place that will be the basis for the business. You have done your homework regarding market, marketing, sales, and production. You have also put a financial plan together to support your analysis (this will be discussed later, for now lets just accept that as part of your business plan, the financials were done.) Aside from the dates mentioned, no other dates are important at this time, except to note that all this activity takes place in the month of January.

Activities

1. The business was organized on January 1 with $20,000 of your cash. It is organized as an "S" corporation.
2. It requires $10,000 of widgets inventory (@$1.00 per widget) and $2,000 of paint inventory. Inventory is purchased on January 2nd on 45-day credit terms, which has been personally guaranteed by you.
3. Equipment needed consists of racks for inventory, painting equipment, and a drying booth. This equipment costs $3,000 (installed); and is purchased for cash.
4. Space requirements are for office, warehouse, inventory, production, shipping, and receiving. Shipping is via a package shipment company. Inventory is

received freight prepaid; and, as you are converting the material for further production, there are no sales or use taxes. Rent is $500.00 per month.

5. You have one sale on January 20th for $8,000 for 4,000 widgets; terms net 30 days.

6. You have hired one employee to do everything regarding receiving and shipping, inventory and production at $10/hr. Payroll taxes: company portion is at a rate of 10% of the gross payroll, the employee portion of payroll taxes (withheld) are also 10% plus the Federal Withholding Income Tax (total 15%) are payable at the end of the month following payroll. You have decided to handle everything else (sales and marketing, all office activities.)

7. You incur other expenses for telephone (on credit) $100, advertising (on credit) $150, office supplies (cash) $200 and rent (cash) $500 (as per number 4 above).

Basic Accounting

We will use the basic chart of accounts discussed previously and create accounts within each major group as we proceed. We have created this initial chart of accounts based upon the activities within the first month. We will create the accounting entries and produce financial statements for the first month.

Chart of Accounts The Widget Corporation	
Assets	
1010	Cash
1030	Accounts receivable
1050	Inventory
1110	Machinery & Equipment
1115	Accumulated depreciation, M & E
Liabilities	
2010	Accounts payable
2020	Payroll taxes payable
Equity	
3010	Common stock
Income	
4010	Gross Sales
Costs and Expenses	
5010	Material cost of Good sold
5100	Direct labor
5150	Payroll taxes
6010	Telephone
6020	Advertising
7050	Office Supplies
7060	Rent
7110	Depreciation expense

Account No.	Description	Debit	Credit
Entries from the above January Activities:			
a. Asset, CASH increases by $20,000, and you now have $20,000 of equity (stock ownership)			
1010	Cash	20,000	
3010	Stock (ownership) to record the sale of stock		20,000
b. Purchase of inventory			
1050	Inventory	12,000	
2010	Accounts Payable to purchase raw Widgets and paint (on credit)		12,000
c. Purchase of machinery and equipment for cash			
1110	Machinery and equipment	3,000	
1010	Cash to record the purchase of machinery and equipment		3,000
d. No entries result from this item			
e. One sale			
1030	Accounts receivable	8,000	
4000	Sales credit sale		8,000
5010	Cost of good sold, widgets	4,800	
1050	Paint from Inventory		800
1050	Inventory: widgets to record the cost ot sale and to relieve for materials		4,000
f. One laborer, four weeks at 40 hours per week			
5100	Labor	1,600	
1010	Net Pay: Cash		1,360
2020	Payroll Taxes Payable		240
5150	Payroll taxes (company portion)	160	
2020	Payroll taxes payable to record monthly labor payroll and the detail of payroll taxes		160
g. Selling and general administrative expenses			
6010	Telephone	100	
6020	Advertising	150	
2010	Accounts Payable to record expenses purchased on credit		250
7050	Office Supplies	200	
1010	Cash		200
7060	Rent	500	
1010	Cash to record expenses paid in cash		500

There is one more entry that is to be made, an adjustment or charge for depreciation. Depreciation is a **non-cash** expense that represents an estimated engineering usage, or reduction in economic value, or Internal Revenue Service (IRS) scheduled (per formula) charge for the wear and tear on the equipment. This will be discussed in more detail in Chapter Five. If the equipment has a "life" of 5 years/60 months then the charge is 1/60th or $50 per month (1/60th of $3000.00). This is calculated on a "straight-line" basis, one of the methods allowed.

Entry			
7110	Depreciation Expense	50	
1115	Accumulated Depreciation to record depreciation expense for the month of _________		50

Creating the Ledger

Now summarize the above entries by account ("posting" each entry to its account); thus creating the general ledger. The source details what created the item in the account.

Account	Source	Amount	
		Debit	**Credit**
1010 Cash	Opening balance	0	
	Deposit	20,000	
	Purchase equipment		3,000
	Pay labor		1,360
	Office supplies		200
	Rent		500
Account balance		14,940	
1030 Accounts Receivable			
	Sale	8,000	
Account balance		8,000	
1050 Inventory: widgets and paint			
	Purchase	12,000	
	cost of goods sold		4,800
Account balance		7,200	
1110 Machinery & equipment		3,000	
	Cash	3,000	
Account balance			
1115 Accumulated depreciation			
	Month ________		50
Account balance			50
2010 Accounts payable			
	Inventory		12,000
	Telephone		100
	Advertising		150
Account balance			12,250
2020 Payroll taxes payable			
	FICA taxes; company portion		160
	Payroll withholding; employee		80
	Payroll taxes; employee		160
Account balance			400
3010 Common Stock			20,000
4000 Sales			
Accounts receivable			8,000
5010 Cost of Goods Sold			
Inventory		4,800	
5100 Labor			
Payroll/cash		1,600	
5150 Payroll taxes			
Payroll taxes payable		160	
6010 Telephone			
Accounts payable		100	
6020 Advertising			
Accounts payable		150	
7050 Office supplies			
	Cash disbursements		200
7060 Rent			
Cash disbursements			500
7110 Depreciation expense			
	Accumulated depreciation		50

Harvey Goldstein

Trial Balance

A Trial Balance summarizes all the account balances to ensure that both sides are (equal) in balance. In addition, a review of the balances may reveal an unreasonable number to be investigated. For example, if sales were $8,050, we would investigate to ensure that the accumulated depreciation offset to the depreciation expense charge was actually made correctly.

Here is the Trial Balance:

The Widget Company Trial Balance As of January 31, 2xxx		
Account	**Debit**	**Credit**
1010 Cash	14,940	
1030 Accounts receivable	8,000	
1050 Inventory	7,200	
1110 Machinery & Equipment	3,000	
1115 Accumulated depreciation		50
2010 Accounts payable		12,250
2020 Payroll taxes payable		400
3010 Common Stock		20,000
4000 Sales		8,000
5010 Cost of goods sold	4,800	
5100 Labor	1,600	
5150 Payroll taxes	160	
6010 Telephone	100	
6020 Advertising	150	
7050 Office supplies	200	
7060 Rent	500	
7110 Depreciation expense	50	
TOTALS	40,700	40,700

Yes, the accounts are in balance as they equal each other. That does not mean that they have all been properly classified. It means that for the time being, we can strike financial statements, and at the least, examine it for reasonableness. We will also be reconciling the major accounts, which include Cash, Accounts Receivable, Accounts Payable, and Inventory to ensure they are correct.

We can now produce Financial Statements:

<table>
<tr><td colspan="3">The Widget Company
Anytown, State
Balance Sheet
As of January 31, 2xxx</td></tr>
<tr><td>Assets</td><td></td><td></td></tr>
<tr><td>Current Assets</td><td></td><td></td></tr>
<tr><td>Cash</td><td></td><td>$14,940</td></tr>
<tr><td>Accounts receivable</td><td></td><td>8,000</td></tr>
<tr><td>Inventory</td><td></td><td>7,200</td></tr>
<tr><td align="right">Total Current Assets</td><td></td><td>30,140</td></tr>
<tr><td></td><td></td><td></td></tr>
<tr><td>Fixed Assets</td><td></td><td></td></tr>
<tr><td>Machinery & Equipment</td><td>$3,000</td><td></td></tr>
<tr><td>Less: Accumulated depreciation</td><td>(50)</td><td></td></tr>
<tr><td align="right">Net Machine & Equipment</td><td></td><td>2,950</td></tr>
<tr><td></td><td></td><td></td></tr>
<tr><td align="center">Total Assets</td><td></td><td>$33,090</td></tr>
<tr><td></td><td></td><td></td></tr>
<tr><td>Liabilities</td><td></td><td></td></tr>
<tr><td>Current Liabilities</td><td></td><td></td></tr>
<tr><td>Accounts payable</td><td></td><td>$12,250</td></tr>
<tr><td>Payroll taxes payable</td><td></td><td>400</td></tr>
<tr><td align="right">Total Current Liabilities</td><td></td><td>12,650</td></tr>
<tr><td></td><td></td><td></td></tr>
<tr><td>Long term liabilities</td><td></td><td>$12,650</td></tr>
<tr><td align="center">Total Liabilities</td><td></td><td></td></tr>
<tr><td></td><td></td><td></td></tr>
<tr><td>Stockholders' Equity</td><td></td><td></td></tr>
<tr><td>Common Stock</td><td></td><td>20,000</td></tr>
<tr><td>Retained earnings</td><td></td><td>0</td></tr>
<tr><td>Current year earnings</td><td></td><td>440</td></tr>
<tr><td></td><td></td><td></td></tr>
<tr><td align="center">Total Stockholders' Equity</td><td></td><td>$20,440</td></tr>
<tr><td></td><td></td><td></td></tr>
<tr><td align="center">Total Liability and Equity</td><td></td><td>$33,090</td></tr>
</table>

The Widget Company Anytown, State For the month ended January 31, 2xxx Income Statement	
Sales	8,000
Cost of goods sold	
Material	4,800
Labor	1,600
Payroll taxes	160
Cost of goods sold	6,560
Gross Profit	1,440
Selling General & Administrative Expenses	
Telephone	100
Advertising	150
Office supplies	200
Rent	500
Depreciation	50
Total S, G & A	1,000
Operating Income	440
Other Income & Deductions	0
Pre-tax Earnings	440

Yes, certain items have been left out, provision for income taxes (an "expense" below the pre-tax line) and the offsetting liability, income taxes payable. However, I have also included both the company's portion of payroll taxes and the employee's payroll taxes, including his income tax withholding. This is to reflect some reality and to demonstrate how it is actually handled.

Again, this was an exercise to demonstrate how accounting operates and how it flows within the accounting system to result in financial statements. No percentages have been calculated, nor have comparative numbers been presented (to prior year or to a budget). These will be discussed in a subsequent chapter.

Summary

1. Accounting is based upon a "double-entry" system in which all transactions are two sided and balanced.

2. A Chart of Accounts is created to organize these entries to produce financial reports.

3. Accounts are categorized into Balance Sheet accounts and Income Statement accounts. Generally, Balance Sheet accounts will always represent value as of that point in time. For example, the accounts receivable balance represents what customers owe the company; and will be collected (real value) or the accounts payable account will reflect how much will have to be disbursed (real value). On the otherhand, the Income Statement accounts generally reveal history that has already occurred, but has no current actual value. For example, sales details how much customers have purchased from the company; but itself will not convert into any current value. The same with advertising (an expense), how much was expended for it?

4. Entries are made into the books. These are incorporated into a (general) ledger detailing, by account, the source of each item appearing in each account. These entries are summarized into a "trial balance" to ensure that all is in balance. The result is to produce two basic financial statements: (1) a Balance Sheet, which states what the company owns, owes, and its net worth as of a point in time, and (2) an Income Statement, which reflects the financial performance of the company over a period of time.

Just a brief note regarding computerized accounting packages: There are many different systems. Each has a system of doing accounting for business in general, for a group of businesses, or designed for a specific industry. Whichever is in use, there are two methods by which to operate.

The first is to "close the books" with every entry: Real Time. Essentially, after any entry is made into the system, all the accounts are summed, the general ledger is complete as a trial balance, and statements are prepared—ready to be printed. The only step is to designate the time period and print the statements either on screen or in actuality. The other system is a "batch" system. In this system, all entries are posted to the system and passes through the particular accounts designated, but nothing else is done until the system is instructed to close in some specified sequence. When the instructions are given, the system then closes with the statements produced.

CHAPTER FIVE —
DEPRECIATION AND TAXES

Depreciation is a NON-CASH expense. Meaning, that the Income Statement will show a charge for this item, which reduces the earnings. However, there will be no actual cash disbursed at any time for this expense. It is an accounting entry that is the result of the purchase of an asset. The cash is expended when the purchase is made. A company buys an asset for $100,000. It disburses the money and then has a depreciable asset on its balance sheet. At that time, no operating cost has taken place, just the exchange of cash for an asset. Examples of assets are: buildings, equipment, automobiles, computer equipment, computer software, building leasehold improvements, etc. It does not include the purchase of land, as land never reduces in value resulting from use. It is always there as is. However, improvements to the land, such as roads, sewers, and water lines do depreciate.

The depreciation results from the use and, thereby, reduction of its worth as an asset. If the asset is used, its ability to perform is reduced. This use or reduction is due to actual use, engineering use, estimated obsolescence, and legal allowances. The "legal" comes from the tax code, created by the Internal Revenue Service (IRS) per the tax laws created by the U. S. Congress. It will outline the methods to be used and the length of time the depreciation can be taken for each type of asset.

There are several types of depreciation methods which are allowed: per unit, straight line, accelerated, and modified accelerated depreciation.

Asset: original cost is $100,000.

Per Unit: The equipment or item is designed to only last through the production of "X" units. If, for example, 100,000 units, then the cost of the equipment will be depreciated 1/100,000, for each unit produced. If the equipment originally cost $100,000, and if in one year the equipment produces 35,000 units, then the depreciation charge will be $35,000.

Straight Line: This is the most common method of all. It assumes that the asset will last a certain amount of time and will be "used up" evenly. If the time period is 5 years for that asset, the depreciation charge will be $20,000 per year without regard to its actual use ($100,000/5 = Depreciation).

Accelerated depreciation: This is mainly used for tax purposes. It is called Accelerated Cost Recovery System (ACRS). Most companies will use the straight line method for internal reporting purposes and their tax filing will be converted to use the ACRS method. It is a formula-based method wherein the earlier years will experience a greater depreciation charge, and the later years a lesser charge when compared to the straight line method. Since depreciation is a benefit when it is used to calculate taxes, this is a preferred method since the company receives greater benefits sooner than later. "Double-Declining Balance" is the formula. Below is the calculation:

Accelerated Depreciation Calculation			
Double Declining Method **Tax Life: 5 Years**			
Since in a straight line calculation, 20% would be depreciated each year, 40% is the factor. However, it is 40% of the declining balance, not the original cost. It continues until the amount falls below the straight line method amount. It then remains stable for that amount until the end of the tax life.			
Installed Cost:		$	100,000
	Percent	Amount	Balance
Year 1	40	40,000	60,000
Year 2	40	24,000	36,000
Year 3	40	14,400	21,600
Year 4			10,800
Year 5			10,800

Modified accelerated depreciation (MACRS): This is similar to the accelerated method, but it is essentially "1.5 Declining Balance" and is used in the same way.

The accounting for depreciation is as follows:

Item: $100,000 piece of equipment, 5 year life; each monthly straight line charge is $1,666.67 ($100,000 I 5 I 12).

		Debit	Credit
Date, monthly	Depreciation Expense	$1,666.67	
	Accumulated Depreciation: Equipment		$1,666.67

Result: The depreciation expense is charged to the Income Statement as an expense, thereby reducing profits. The offset (credit) side of the entry is to Accumulated Depreciation. It is a "contra" asset account. The term "contra" is used to describe a situation where an account (such as an asset account) is normally expected to have a debit (or left-side) side balances; but has a right side balance. Remember the Chart of Accounts!!!

Value of the asset is as follows, Balance Sheet Accounts

	Debit	Credit
Equipment Asset Account	$100,000	
Accumulated Depreciation		$1,667
Thus, the net book value is:	$98,333	

The value of the asset on the company's books has been reduced. The asset account is reduced by the contra (accumulated depreciation account) to result in to the "net book value."

Reiterating, under our tax laws, the purchase of an asset can not be used as an expense against taxes as it lasts for a period longer than one year. Thus, the cost of such a purchase is spread out over its "life." The depreciation represents the cost of that asset for that period of time. The benefit is that it reduces the taxes owed while not being an actual cash disbursement. The cash disbursement was made when the asset was purchased. See the example below.

<table>
<tr><td colspan="3">XYZ Corporation
Income Statement
For the Period Ended December 31, 2xxx</td></tr>
<tr><td></td><td>Without Depreciation</td><td>With Depreciation</td></tr>
<tr><td>Net Sales</td><td>$ 1,000,000</td><td>$1,000,000</td></tr>
<tr><td colspan="3">Cost of goods sold</td></tr>
<tr><td>Material, Labor and Plant Exp</td><td>600,000</td><td>600,000</td></tr>
<tr><td style="text-align:right">Depreciation</td><td></td><td>50,000</td></tr>
<tr><td>Total COGS</td><td>600,000</td><td>650,000</td></tr>
<tr><td>Gross Profit</td><td>400,000</td><td>350,000</td></tr>
<tr><td>Selling, Gen'l & Administrative</td><td>150,000</td><td>150,000</td></tr>
<tr><td>Pre tax Earnings</td><td>250,000</td><td>200,000</td></tr>
<tr><td>Income Tax @ 40%</td><td>100,000</td><td>80,000</td></tr>
<tr><td>Net Income</td><td>150,000</td><td>120,000</td></tr>
<tr><td></td><td>(higher)</td><td>(lower)</td></tr>
<tr><td>Add Back Depreciation*</td><td>0</td><td>50,000</td></tr>
<tr><td>Cash Net Income</td><td>150,000</td><td>170,000</td></tr>
</table>

*For cash calculation purposes, depreciation is added back as no actual cash was expended.

Thus, without depreciation, the earnings were more by $30,000. However, the affect on income taxes with the depreciation was to reduce the earnings by $20,000 (40% is the tax rate on the earnings of the $50,000), which resulted in having $20,000 more in cash. All else was the same, just the "non-cash" expense was charged to the statement, reducing the taxes (cash sent to the government) and thus, increasing the cash.

Certain items are AMORTIZED. This, too, is a non-cash expense. It is similar to depreciation, but for other than tangible assets. For example, let's take the "Goodwill" account. It is the excess of the purchase price paid for the assets over the amount that can be specifically applied to the assets purchased. This excess, is "written-off" over fifteen years; just as deprecation is expensed. However, the amortization period is longer than most depreciation periods. Thus, each amortization period charge amount is lower and less beneficial from a cash basis. The accounting affects on the Income Statement and the Balance Sheet are the same as with depreciation (Goodwill less Accumulated Amortization.)

In summary, depreciation and amortization are non-cash charges that reduce earnings, and thus, income taxes to be actually paid. The result is an improved cash position. The impact on the Balance Sheet is to reduce the asset balance via the offset to the depreciation expense accounting entry by crediting the "accumulated" account. For depreciation, the actual expenditure of funds is in an initial period, purchasing an asset. For amortization, it may be from several sources; such as, goodwill, organization expense (also capitalized, creating, and asset account), patent expenditures, etc. These are actual expenditures for other than tangible assets.

Taxes (on Income)

Income taxes are paid by businesses to various governmental entities based upon earnings. Different legal forms of business (see Chapter Six, Legal Structures [and Tax Basis]) will pay these taxes via various methods. However, the tax liability for each company and its ownership will be satisfied.

A "C" corporation is one that usually has in excess of 35 owners. Further, "C" corporation taxes are calculated upon the tax laws (certain expenses may not be fully tax deductible as expenses against earnings), including depreciation method used. Below is a tax payment schedule for such a business. The tax schedule is "graduated," (i.e. the more the company earns, the greater the tax percent is owed for taxes). This schedule is in effect as of January 2000.

U.S. Corporate Tax Rates as of January 2000		
Taxable Income Brackets	Amount of Tax For Base of Bracket	Plus this percent on excess over the Base
up to $50,000	0	15%
$50,000 - $75,000	7,500	25%
$75,000- $100,000	13,750	34%
$100,000 - $335,000	22,250	39%
$335,000 - $10,000,000	113,900	34%
$10,000,000 - $15,000,000	3,400,000	35%
$15,000,000 - $18,333,333	5,150,000	38%
Over $18,333,333	6,416,667	35%

1. To calculate your overall tax rate, take the calculated tax and divide by the pretax earnings.
2. The "bracket" (marginal tax percent) is that percentage in the right column that is the last level that the earnings fall into, that is, the last percent used.

For example: Taxable earnings are $12,500,000.

The tax is calculated as follows: $3,400,000 (the tax on the base bracket amount of $10,000,000) plus 35% of $2,500,000 ($875,000). Total tax liability = $4,275,000

CHAPTER SIX —
LEGAL STRUCTURE (AND TAX BASIS)

This section will briefly discuss the basic legal structures of various economic entities (companies) and the advantages and disadvantages of each.

If you are going into business or are already in business, it would be wise to consult an attorney on this subject. This section may provide some insight as to which status is best for you, but no more than that.

There are three basic legal structures: a) sole proprietorship, b) partnership, and c) corporation. In addition, there are several varieties and combinations of each, which provides more flexibility and alternatives.

Sole Proprietorship

The basic structure is the individual(s) who owns a business. The proprietor or owner and the business are one and the same. That is, a mom and pop grocery store; wherein, they are the store and the store is them. They file the normal IRS Form 1040. However, they will not be completing the Salary and Wages line on the front page of the form, but an SE (Self-Employed) Form. The SE Form (Schedule C) details and itemizes all revenues and expenses associated with their business, with the "profit" (or "loss") being carried to the proper line on the face page of the 1040 form.

With a sole proprietorship, one can "open the door in the morning" if the owner wants to or not. It is only taxed once. (Schedule C becomes part of the overall individual tax filing.) However, any legal liability that may occur belongs to the owner or sole proprietor. Again, so do the profits. The owner must arrange for everything including all insurances. This is the simplest form of legal ownership.

Partnerships

A partnership is a legal business entity wherein there are several owners. No incorporation is made. Each has invested something into the business. There are different types of partners.

A partnership has a limited life as any change in the partnership will cause a dissolution.

1. <u>General Partner</u>: This partner has invested something and been awarded some portion of ownership. Most common is an investment of cash. It could also be "know-how" or the ability to bring business to the business. This partner is active in directing the business. He enjoys the profits of the larger business and pays taxes on the dividends declared and paid. However, he is also liable for any act the company may incur, regardless of his investment proportion.
To expand, if partner A invests $100,000 of a total business investment of $1,000,000, he owns 10%. He is also entitled to 10% of the earnings. However, he may and can be at risk for most of any liability. For example, if an action occurs where the company is sued and loses the action for $2,000,000, the partners are then liable to pay (let's assume no insurance coverage is in effect). In a normal situation, he would have to contribute 10%, or $200,000 as his portion. However, his risk doesn't stop there. The total net worth of all the partners are at risk (unlimited liability). If the other partners only have $1,500,000 amongst them, and partner A is worth $3,000 000, then he could be liable for the $500,000 versus his proportional $200,000.

2. <u>Limited Partner</u>: This individual has invested a sum of money in the business for pure investment. The partner is not active in any aspect of management. This partner is eligible for a proportion of the earnings. However, his liability is limited to only his total investment, as he, again, is not active in the company.

3. <u>Managing Partner:</u> This partner is the actual head of the organization; the CEO of the partnership. Aside from this operational status, there is no difference in ownership or liability from a general partner.

There are several variations of a partnership entity. The Limited Liability Partnership (LLP) has the advantages of limited liability (as in a corporation, see below) and enjoys the tax advantages of a partnership. A Professional Corporation (PC) is one wherein the individual partners may be individually liable for their own actions. Thus, unlimited partnership liability is eliminated; similar to a corporation.

Taxes are paid on their share of company earnings. It is paid just the one time. The payment is based upon their personal total personal tax returns.

Corporation

This is a legal entity or being. It is generally described as a "C" corporation. Initially, every such organization required an Act of Congress. As corporations popularity increased, the States were given the power to create such corporations and now a simple application is required to create a "C" corporation. The Company's Charter and Bylaws outline its character and goals. It is a legal person that never dies, unless (1) it declares bankruptcy and is dissolved; (2) it is purchased, merged, etc. with another company and loses its own identity; or (3) is expanded as it has acquired the other business. Ownership is in the form of shares of stock. Percentage ownership is via the number of shares owned versus the total number of issued and outstanding shares.

Since the corporation is an entity itself via its incorporation, it retains its own legal liability and protections, and pays its own taxes. Thus, if there is a suit against the company, and if it loses all of its assets, the value of the company is zero and so, generally, is its stock value. However, that is all the owners or stockholders can lose. Their other personal assets cannot be affected. There is nothing else at risk, just their investment. The most they have at risk is that the value of their shares declines to zero. Compare this to a partnership or a sole proprietorship.

Corporate taxes also present a different picture. An entity or company does business and makes a profit. According to the tax schedule, after all tax shelters are taken, the income tax is computed and paid. This is the "first" income tax paid. The dividends are calculated and distributed to the stockholders, generally upon the Net (after-tax) earnings. The stockholders report their dividends on their own personal income tax filing, and they pay taxes on it. Thus, the company's earnings are taxed twice; once when they are earned by the company and a second time by the recipients of the dividends.

Sub-chapter S corporations and LLCs

These are entities that are similar in nature. They have legal protections as a corporation, but are only taxed once. Basically, they have no more than 35 owners and must operate as a corporation with, at a minimum, annual board meetings, board resolutions, minutes of meetings, etc. Thus, they enjoy all the legal protections of a corporation, but the owners only pay taxes on their portion of the pre-tax earnings (or a tax shelter, if a loss year). However, there is one catch. Remember, a profit may not be a cash profit. Thus, if no dividends are declared, the owners are paying taxes on profits of the Sub-chapter S corporation without the company distributing any cash that would cover the extra taxes they must pay on their portion of the earnings. Therefore, if a sub-chapter S corporation makes a profit, but does not distribute dividends to its owners, the owners must pay the taxes on their portion of the profit with their own cash. This is a major consideration!

Let me reiterate, this should be discussed with legal and tax professionals.

SECTION TWO

Understanding and Planning

CHAPTER SEVEN — WORKING CAPITAL

The management of working capital is a major function of the Finance Department and of any manager or entrepreneur. This chapter defines Working Capital and analyzes potential policies.

There are two definitions of Working Capital: Gross Working Capital and Net Working Capital. Gross Working Capital consists only of a company's current assets, which consist of cash marketable securities, accounts receivable, inventory, and any other current assets. Net working capital subtracts the current liabilities from the current assets, as they also must be paid within a similar time frame. Current liabilities are accounts payable, accrued payables and current liabilities for all taxes, payroll, and short-term borrowings (such as line of credit, loans from officers, current portions of long term debts). Current means one year or less.

The management of working capital is a major function of the finance department; or of any manager or entrepreneur.

Current Assets

Cash: How much does the company need? A company needs an amount of cash to cover normal operating disbursements. Some extra cash to cover emergencies and perhaps an additional amount to take advantage of special situations, which are of great benefit to the company. Or, possibly the company wants to carry as little as possible to enhance the owners' cash or reduce their asset basis. This will improve the return on assets (net earnings divided by gross assets). Reduced cash, thus reduced assets, thus reduced investment base would generate a higher return on investment. It could also leave a company in a tough position should business slowdown.

Chapter Twelve will discuss cash flow in more detail—what cash flow is, how it is measured, how cash flow is planned annually, monthly, and shorter projections for a 30 to 60 day period.

Accounts Receivable (A/R): Accounts receivable starts with the policy set by your industry, the policy of your company, and the enforcement of that policy. It initiates with the general terms and conditions under which a customer is sold. Your industry's generally accepted "rule of thumb" may be, for example, 2% net 30 days. This does not dictate your company's policy; but the customer will, most likely, expect it. However, your company may adjust these terms by accepting payments without consequence through 60 days. You may choose to change the discount to 1% or 3%, make the discount days 10th prox, or 15 days, etc. You may allow a discount to be taken up to 5 days after the discount period ends. You may call for money on the 31st day or the 45th day or the 55th day. Those are just a few of the items that will impact your company's A/R policy and the total A/R level working capital amount and cash flow.

The A/R policy and enforcement of it has several impacts:

1. How "tough" will the company be to do business with. This includes refusal to continue to do business with the customer until he returns to acceptable payment terms. After all, sales are difficult to secure. However, if payment is not forthcoming, then only the sale will cost money (the costs have already been recorded on the income statement) will be written off as a bad debt. Costs also include the product and all the administrative expense to collect. It is assumed that the product will not be returned. However, if it's worthwhile; it may be repossessed if it's still in your customer's possession.

2. What portion of the company's receivables will be in a position to support a loan should it be required? Many banks use receivables to secure short-term loans.

3. How much monetary exposure does the company want to have out in the market? The greater number of days outstanding (DSO), the less likely or

more difficult it will be to collect. The DSO is the average number of days it takes to collect the credit sales (see the formula below.)

4. How aggressive does the company want to be regarding securing business via the credit function? This could mean approving credit for lesser credit worthy customers or higher credit limits with current customers. The easier a company's credit policy, the more sales a company can secure. But, how prompt will collections be for the customers and how many will be uncollectible?

5. What is the cost of the money not collected? This includes the alternate costs of money and the actual cash not in the company's coffers. For each day the average collection period (or DSO) can be reduced, via enforcement or reduction of float, one must assume a worth of at least 5% per annum rate. This assumes that if you had the money by collapsing the DSO, the money would be invested in short-term investments. Other lost alternatives include: (1) the discounting of certain payables at possibly very handsome rates of interest (see the section on accounts payable below in this chapter); (2) repayment of other short-term loans; (3) having a cushion in the bank, which your company may want; (4) invest in longer term investments, including securities and other company investment (higher returns); or (5) the ability to distribute higher dividends to shareholders. Each day of reduced outstanding collections represents an average day's sales/collections in your company's bank account.

The following is an example of reduced DSO:

To calculate this, divide the total sales by the operating days in which collection and banking is performed. This could probably total about 250 banking days (fifty-two weeks at five days per week less ten days of company holidays). Thus, if your company has annual sales of $250,000,000, then the daily collections is $1,000,000 per day. At 5% per annum, the additional collection will earn an additional $50,000 if invested in short-term securities. Many companies use a 360 or 365 day year. It too, generates important cash.

If you regularly offer terms (a discount for prompt payment), it can be quite advantageous for customers to pay promptly. This may be a tool for earlier collection if desired or emphasized with certain customers. Again, the value will be detailed in the accounts payable section below.

In item 5 above (cost of money), we alluded to the DSO number to measure the average collection period. There are several methods to calculate this DSO number. The one, I believe, best illustrates the number of days is to divide the month-end accounts receivable balance by the sales total that occurred within the particular month, which is one of the causes of the month-end balance. Then, multiply by the number of days in that month. Note that this assumes that all sales are on credit. COD (Cash On Delivery) and CIA (Cash In Advance) should be so small as to be inconsequential, and cash sales should not be material (under 10% of the total sales). If the cash sales are more than 5%, the sales data in the formula below should represent just the credit sales.

Formula: DSO =($Receivables / $Sales) x days in the month being reviewed

If there are accounts receivable of $1,000,000 as of June 30, June sales of $600,000 with thirty days in June, then the DSO would be ($1,000,000 I $600,000) x 30 = 50 days outstanding in collections, or a $1,000,000 A/R balance as of February 28th for sales in February of $558,000. Then ($1,000,000 I $558,000) x 28 = 50.2 days.

Utilize this formula to determine an annual DSO number of days, simply average the last twelve months or more (up to three years). Beyond three years, business has changed and the conditions are probably not the same: economy, product, market, etc.

Further, if a company wants to track the monthly DSO data, a trend may be detected. Another method, which may be beneficial, is to track on a three-month or six-month rolling average. In this method, an average DSO for the periods studied can be calculated and as a new period is calculated and included into the average, the oldest period is dropped from the calculations. This method moderates any one-month quirk. (A quirk? Periodically, a particular month may experience a funny, but accurate, number for any one of a multitude of reasons.)

Summarizing Accounts Receivable: A liberal credit policy may increase sales but cause slower collections or greater bad debts. It may also affect your ability to borrow on an A/R collateralized line of credit, as most banks will support a loan with approximately 75% of those receivables under 60 days in aging. If the company increased its cash requirements via increased sales, it may not be able to collateralize a revolving loan due to slower collections and lower quality receivables. It may also raise the company's credit requirements above its line limit.

A tight credit policy may cause reduced sales, lower or slow production, thus causing your unit costs to increase. Customers may look elsewhere as your company may be perceived as being too hard to do business with. Supply and demand or product uniqueness may put your company in a stronger or weaker credit policy position.

Another current asset is Inventories: This, too, has the following interesting implications:

1. How much inventory is enough to support sales? What type of inventory is necessary? Raw inventory, in-process inventory, or finished goods? Delivery time and safety stock. As a "rule of thumb", it is assumed the cost of carrying inventory is between 20% and 25% per year of the inventory value. This includes the monetary interest cost of carrying the inventory, shrinkage, damage, and cost of excess labor to maintain the large inventory. Thus, why have a lot of inventory? Because the company may want to have the ability to service customers at a whim; or if inventory is insufficient, it may cause shipping delays and dissatisfied customers. Additional factors influencing the amount of inventory necessary are the time to convert to finished product and the internal assortment of products.

2. Obsolescence and physical disappearance are real concerns also.

3. The ability to support potential business loans. If the inventory is accepted as collateral for a loan, it may only be valued at 50% of its worth. From a bank's perspective, it is not an immediate source of cash. It may have to be converted into a salable product. Also, it may have a limited market and, in a crunch situation, potential buyers will offer a reduced price to purchase it.

If you have a purchasing department, they should execute your inventory policy.

Summarizing Inventory: Policy consideration should be given regarding the availability of inventory for sales versus obsolescence and shrinkage. If it is decided to always have more than enough to service the customer; carrying costs come into play, in addition to the obsolescence and shrinkage, which are real costs. Carrying costs include warehouse maintenance, labor, and cost of loans, are needed to support the inventory. Also, what would the company do if there were no inventory excess to carry, et al. Repeating, the "rule of thumb" is that the cost is estimated to be between 20% and 25% of the inventory value annually. If a tight policy is decided upon, then the ability to produce and ship within a reasonable time frame may not be something that can be accomplished. In addition, a reduced labor force may result that could be of insufficient size to handle a rush of business, or time to train new employees. This, also, must be considered. What is tight and what is liberal can be measured by the days in inventory calculation (inventory I average daily cost of sales). See Chapter 8. Whatever policy is chosen, inventory turnover will be affected.

Marketable Securities: This must be viewed as a temporary situation. It is not the goal of the business to be in the investing business, unless it is so stated in either a business plan, or is an industry necessity (as with insurance companies or securities companies); or there are no other avenues for the company to earn a profit other than investing for a period of time. If this is the case, an acquisition may be a possible use of these "excess funds," or the company could declare an extra dividend and distribute the extra funds to the owners. Three caveats: (1) review the cash needs on a longer basis (via a twelve-month financial plan and create a strategic plan where the cash is a necessity); (2) if other investors become aware of the funds, it may attract a "raid" (a hostile takeover bid to eventually gain control of the cash); and (3) from a tax standpoint, the IRS may view an extended length of time that a company holds "excess" cash or marketable securities as tax avoidance (depending upon the legal structure). To maintain such high amounts will enhance the stock value with long-term capital gains, which is taxed at lower rates. However, if the excess cash were distributed as dividends, then the IRS would be collecting taxes sooner from the recipients at current (higher) rates.

Current Liabilities

Accounts Payable (A/P): It is also subject to decision and policy. Does your company prefer to pay within the discount period whenever one is offered; or, per policy (or cash demands), stretch the payment as long as possible? This can be done per overall policy or selective policy.

It is important to detail the value of taking a discount. Example: Normal terms in your industry may be "2% 10, net 30." This means if your company pays the invoice within

10 days of receipt of the good or service, it can reduce the amount that has to be paid by 2% (this usually does not include any discount on freight). Sales and Use taxes, if any, will also be reduced as the product total declines. What is this worth to your company? The formula to calculate this is:

%value / interest= (discount offered / 100%-% discount offered) x
(365/# days from end of discount period to net date)

Stated algebraically: % opportunity cost = $\dfrac{a}{1-a}$ **X** $\dfrac{360}{c-b}$

Where, a = discount offered

b= discount days

c= days after discount period to net pay date.

Thus, if the terms of sale offered by a vendor are 2% 10, net 30, then in this example:

% interest = $\dfrac{.02}{.98}$ x $\dfrac{365}{20}$

% = 37.2

What does this mean? It means that payment of invoices within (best on the 10th day) the discount period will yield at an annual interest rate of 37.2%. Where can someone or businesses get that kind of return? It's difficult to find.

Harvey Goldstein

Some other common discounts offered:

% annual interest

1% 10 net 30	18.4%
2% 15 net 30	49.7%
2% 20th, net 30	74.5%
2% 20th, net 60	18.6%

Summarizing Accounts Payable and Policy considerations: Does the company want to stretch the payments as far as it can go? This will maximize the cash flow. However, your vendors may put your orders in a lower priority position because of it, which may affect a rush of business. If you are stretching, and there is a business slow down, and your cash position is weak or you are bumping the top of your Line of Credit, it may affect your ability to pay and affect your credit rating. On the other hand, early payment of invoices may enhance your bottom line via reduced costs and per the discount discussion above. But it may also reduce your cash flow as you are disbursing funds "earlier" than later via stretching.

Other current liabilities include certain legal liability accounts such as payroll taxes and income taxes. Also, any portion of any long-term debt that is due within one year of the statement date is considered current. Any revolving line of credit is considered current as it is payable on demand at any moment, not on a long-term schedule.

Working Capital Summary: A tight asset policy would enable your company to reduce its asset requirement, but it may also put you in a tight position should business slow-down. Examples include:

1. It could impact your ability to move quickly to take advantage of special situations (i.e. special pricing on regularly purchased materials or special large quick customer orders).

2. It could also affect your business level by reducing the number of accounts via the credit function and not having inventory with which to produce product.

3. A smaller inventory requires more exact distribution between all inventory items and timing and forecasting must be raised to a higher level. And the opposite policy, ready for anybody and anything—at a cost.

These decisions can be tested within a financial plan. We'll do this in Chapter 10.

Float

Float is the time taken for payments to clear the banking system. It can either be in your favor or work against you. There are three main areas of float:

- Clearing Float
- Mail Float
- Processing Float

Clearing Float: In its simplest form, clearing float, one writes a check, thus the checkbook shows a reduction in the balance. However, between the mail and the processing of the check by the recipient, it may not clear your bank (the funds actually leaving your account) for five business days. Therefore, you actually have the use of the money for that period of time, while your checkbook shows a reduced balance. In the interim, your account appears to be paid with the vendor who deposited it; but, not yet actually receiving the funds. Obviously, the reverse occurs when a customer pays you. However, if the bank gives you immediate credit for deposited checks, then you have money available to you without the check actually clearing.

Mail Float: To improve the incoming mail float (and clearing float), a lockbox at your collections bank would enhance the situation. The payments are sent directly to your bank and immediately enter the banking system. A copy of the check and all paperwork subsequently is mailed to your office for processing, together with the deposit slip. This saves one to two days for processing and depositing. Also, if your office only deposits once or twice per week, then the days saved become more significant. If there is a significant amount of payments from all over the country, then it may be to your advantage to set up regional banks for collection and processing, saving mail days, etc.

Processing Float: On the other hand, when making payments, your company will pay a regional bank in Seattle from one of your banks, for instance, in Boston; thus, enhancing your float. Your customers are doing the same thing, for the most part. All of this can be done, if it's worthwhile!

The means to accomplish the analysis is to determine your course of action and cost it out. For example, set up a lockbox system and have your bank detail the costs and fees. Estimate how many days in float time you will save. Convert this to cash by multiplying the number of days saved by the average daily collections (annual sales on terms....no cash sales or credit card sales, divided by two hundred and fifty banking days). Two hundred and fifty days is fifty weeks times five work days per week. This also accounts for various holidays the banking system is closed. Then, multiply this collection advance (saved days x dollars per day) by the cost of capital, or the Federal Funds rate or your actual short-term bank borrowing rate. Those are the savings, to measure against the cost. This collection advance is also what your company will receive, initially, as a one-time additional cash flow; as a result of the collapse of the float days.

Example: Annual sales = $3,600,000; average per day (360 day banking year): $10,000.

at 5% per annum rate of interest = interest savings or value of $500.00 per year for each day saved; and you have the actual $10,000.00 in your account.

Note: If you are borrowing via a line of credit at, for example 10%, then a very direct consequence of reducing the collection days would be a reduction of interest expense at the 10%, per annum, for the amount reduced by the improved collection. Or if your company has a variety of funding, various equity, and debt securities, the weighted cost of capital for the company should be used versus the 10% cited in the example. See Chapter Fifteen Investment Analysis: Capital Budgeting; Cost of Capital for this computation.

CHAPTER EIGHT —
FINANCIAL STATEMENT ANALYSIS /
FINANCIAL RATIOS

In this chapter, we will discuss various financial ratios, their usage, and value in the analysis of your company's operations.

We have gone through the financial statements. We know they tell us how we've performed and how much we are worth, on paper. We will make comparisons to other periods and other standards, in another section later. The ratios themselves don't tell a story without being compared to either previous periods, industry standards, general business standards, or standards set by a lending institution or the investment community. One, also, must consider a lending institution (a bank for example) as an investor. The institution would have money placed within the control of the business and its management, as an investor would. It just has a higher "more secured" priority in getting its money back versus an equity investor. Thus, it is also most interested in how well a company is performing, current and in the future. The final comparison is what will satisfy management and ownership considering their own requirements and background.

Let's examine the ratios we will be using and what they tell us. There are three categories of ratios:

- Liquidity
- Operating Efficiencies
- Finance Ratios

Harvey Goldstein

In order to provide additional clarification of the data being presented, a Balance Sheet and Profit and Loss Statement are inserted. Actual example data and calculation will be made from these statements. These statements are from those presented in Chapter Ten, Budgeting and Financial Planning. The data developed from your statements should be compared to your industry "norms" to make relevant comparisons.

XYZ Company
Financial Statements for December and Total Year

<u>Income Statement</u>	<u>December</u>	<u>Total Year</u>
Net Sales	601,048	7,300,000
<u>Cost of Sales</u>		
Direct Materials	409,405	4,972,415
Direct Labor	71,241	865,252
Plant Expenses		
Utilities	4,607	55,279
Depreciation	2,186	26,227
Rent	3,452	41,424
Health Insurance	8,943	107,312
Shop Supplies	1,843	22,381
Payroll Taxes	10,472	127,192
Total Plant Exp	31,502	379,816
Total Cost of Sales	512,148	6,217,482
Gross Profit	88,900	1,082,518
%sales	14.8%	14.8%
<u>General & Administrative Expenses</u>		
Salaries (Incl. Plt. Mgr)	20,702	248,420
Payroll Taxes	2,246	26,951
Commissions	7,160	86,963
Advertising & Promotion		26,000
Auto Expenses	1,500	18,000
Misc Txs & Licenses	1,000	3,500
Office Expense / Postage	2,000	24,000
General Insurance	2,500	32,250
Repairs & Maintenance	100	1,200
Dues & Subscriptions		2,400
Travel, Meals & Entertainment	3,000	36,000
Telephone & Utilities	1,335	16,021
Miscellaneous	200	2,400
Total Gen & Admin	41,743	524,105
%sales	6.9%	7.2%
Income from Operations	47,157	558,413
%sales	7.8%	7.6%
<u>**Other Income & Pedycts**</u>		0
Interest expense	(4,667)	(56,000)
Interest expense	3,005	36,500
Bad Debts		
Total Other Income / Deducts	(7,672)	(92,500)
Pre - tax Income	39,485	465,913
Income Taxes	13,820	163,069
Net Income	25,665	302,843
%sales	4.3%	4.1%

XYZ Company
Financial Statements for December and Total Year

Balance Sheet

Current Assets	as of 12/31/xy	Average 2xxy
Cash	9,650	
Accounts Receivable	1,039,677	
Inventory	565,312	584,378
Due From Officers	36,450	
Total Current Assets	1,651,089	
Plant, Property & Equipment		
Machinery & Equipment	1,181,150	
Office Furniture and Fixtures	75,610	
Total P, P & E	1,256,760	
Less: Accumulated Depreciation	584,927	
Net Plant, Property & Equip	671,833	672,437
Total Assets	2,322,922	2,324,241
Current Liabilities		
Curr port. of Long term debt	88,520	
Accounts Payable - Trade	851,102	
Taxes Payable	52,998	
Line of Credit with Bank	270,959	
Total Current Liabilities	1,263,578	
Long Term Liabilities		
Notes Payable	80,130	
Total Liabilities	1,343,708	
Stockholders' Equity		
Capital Stock	437,480	
Retained Earnings	541,733	
Total Stockholders' Equity	979,213	
Total Liab & Equity	2,322,922	

Key Ratios / Data		Average 2xxy
Net Working Capital	387,511	
Current Ratio	1.31	
Quick Ratio	0.86	
Times Fixed Charges	9.5	9.3
Inventory Turns: YTD projected	10.64	10.64
DSO	53.62	51.19
Asset Turnover	10.9	10.9
Debt Ratio	0.578	0.644
Return on Equity	30.9%	36.6%

Liquidity Ratios reveal a company's ability to meet its obligations (pay its bills).

Current Ratio

<u>Current Assets</u>
Current Liabilities
= Liquidity

Formula: current assets divided by current liabilities equals the current ratio. Example:
Current Ratio = $1,651,089 / $1,263,578 = 1.306

This is a basic measure of a company's liquidity, its ability to meet its current obligations. Those are usually due within 60-90 days, at most, and others (such as a line of credit or the current portion of a long-term debt) that are due within one year. It is assumed that all the current assets (principally cash, accounts receivable, marketable securities, and inventory) can be converted into cash within that period to meet the obligation. Depending upon the "norm" of the industry, an acceptable liquidity ratio may be 1.25:1.00

Quick Ratio (Acid Test Ratio)

<u>Current Assets - Inventory</u>
Current Liabilities
= Liquidity

Formula: current assets less inventory divided by current liabilities:
Quick ratio = ($1,651,089-$565,312) / $1,263,578 = .859

Inventory is eliminated from the assets as it does not convert as quickly into cash. The inventory either has to be converted into product, sold, and collected (potentially a 90 to 120 day period) or it may have to be sold quickly, as such, with a potentially significant loss in value, if a buyer can be quickly identified. Thus, inventory is eliminated from consideration in this calculation.

It demonstrates whether there is, or is not, enough quick liquid assets to quickly satisfy your immediate obligations (current liabilities). The ratio could be 1.00:1.00. However, different industries will have differing ratios; some as low as .5:1.0 or less

Accounts Receivable Turnover

Formula: Credit Sales / Accounts Receivable = Accounts Receivable Turnover
Accounts Receivable Turnover= $7,300,0001/$1,039,677 = 7.02

This is generally reviewed at year-end and usually takes the A/R as of a point in time versus an average over the period. If it is struck at a time other than for the whole year, then adjustments have to be made to produce an annualized number. Reviewing several years of data may reveal a trend, but it is just too much of a snapshot, one time per year. However, if it is a rolling prior twelve month sales (to that date) amount divided by the accounts receivable each month end, then it may provide some information regarding A/R efficiency. That ratio would be: last 12 months credit sales / current A/R. This should be struck each month and charted for seasonality, trends, etc.

Days Outstanding in Receivables (DSO)

This is a more popular analysis of the A/R collection activity. And is the one I, and most credit managers, prefer. It reports the average day's time to collect receivables.

The Formula: Accounts Receivable x   number of days for month just ended
Sales for the month ended
DSO = $1,039,677 / $601,048 = 1.73 times 31 days (in December) = 53.6 days

When this is calculated each month, a trend and average can be calculated. The only caveat is if there is a significant amount of cash (non-credit) sales. If this is the case, then the cash sales must be subtracted from the monthly sales total used in the calculation. The remaining total is the credit sales. Cash sales are collected immediately and credit sales reflect the number of days outstanding per the calculation (and are in the A/R balance). Again, depending upon the industry, it could average 45 days for manufacturing or distribution or 14 days for the dairy industry.

Inventory Turnover

Formula: Cost of Sales divided by average inventory (for the period being analyzed).

The ratio must be adjusted to annualize the results, if it is calculated at any point during the year other than for a full year. This can be accomplished by taking the number of months covered by the period, dividing that by 12 to get the percent of the year covered. Then, taking that percent and dividing the material cost of sales by it will "annualize" the cost of sales. This is to be divided by the average inventory calculated.

<u>Annualizing example:</u>
Month: May: month 5, thus 5/12 = .417
May year to date (YTD) cost of sales $1,200,000 (little seasonal fluctuation).
Therefore: $1,200,000 / .417 equal annualized material cost of sales of $2,877,698.
Five month Average Inventories: $720,000.

Thus, $2,877,698 / $720,000 = 4.00 Annualized Inventory turns

If your business is highly seasonal, then the cumulative seasonality factor should be used to project the annual direct material cost instead of the month factor per the above example. Seasonality will be demonstrated in Section Three.

The result will be the annualized cost of sales projected from that period of the year divided by the average inventory for that period. In the example, the average inventory for the January-May period. One other method is averaging the past 12 months of cost of sales divided by the average inventory for that period. However, as this is not specific for the current fiscal year, then it must be charted and a trend analyzed.

> The Formula: annualized (12months) cost of sales / Average Inventory
> Inventory Turnover= $6,217,482 / 584,378 = 10.63
> (From the sample statements.)

It details how efficiently your inventory is utilized. The higher the turnover number, the less the investment in inventory is required to generate and support the sales. Restated, greater inventory turns means that for the same dollar investment in inventory, greater sales are produced.

Total Asset Turnover

Formula: Sales divided by total assets. As with the Inventory Turnover, this is an annual number used in analysis. Thus, if a calculation is made during a year, an adjustment must be made to annualize the partial year number. This, again, is calculated as the inventory number, above, or twelve months of sales divided by the average monthly assets for that period.

Sales / Total Assets (including current, long term, and other assets)

Total Asset Turnover = $7,300,000 / $2,324,241 = 3.14

It demonstrates how efficiently your company is utilizing its investment.

Fixed Asset Turnover

The Formula: Sales divided by net investment in fixed assets.
Sales / Net Fixed Assets
Fixed Asset Turnover = $7,300,000 / $672,437 = 10.86

This demonstrates the efficiency of the actual productive assets and is the more popular of the asset turnover ratios. It is my preference also. Please, also ensure that an adjustment is made if it is calculated for any point other than for a full year. Again, a company can also utilize selling data: twelve months of sales divided by the average of the monthly net fixed assets for the same period. This is used and particularly analyzed where an industry requires heavy investment in capital equipment.

Operating Ratios, used to detail the efficiency of the company.

These are income statement ratios or earnings percent.

Net Sales is the denominator and is defined as sales after Returns (R), Discounts (D), and Allowances (A) are subtracted from the Gross Sales. Gross Sales are the pure total of all product or service invoices (excluding sales taxes and freight). Subtracting the R, D & A from their gross sales will generate Net Sales values to the company of all items sold.

In some instances, when the company pre-pays and bills for the freight to the customer this, too, is deducted from the gross sales amount to generate net sales of the company's product or service. (In my opinion, freight "sales" and costs should be matched and placed into a cost account area. This is just to present alternatives and the potential impact on the Net Sales value).

Gross Profit Margin

The Formula: net Sales less the cost of sales divided by net sales. Gross Profit is that level of earning only after the direct cost of sales is subtracted from the net sales. It is easily identified as such on the Income Statement. The cost of sales should include direct material, direct labor, production expenses and shipping, and receiving department labor and expenses.

> (Net Sales- Cost of Sales) = Gross Profit / Net Sales
> Gross Profit Margin = ($7,300,000-$6,217,482) = $1,082,518 = 14.8%
> (From the sample statements)

Are the products generating enough profit for the effort to produce them? Either the prices are too low or the production efficiencies inadequate. Again, why?

This is viewed monthly and on a Year-to-Date basis (YTD). YTD is smoother versus a month's snapshot that may contain some odd items. But in all cases, explanations are important.

Operating Ratios: Selling expenses (group of accounts) as a percent of Net Sales

> General and Administrative expenses, percent of Net Sales and any other group of expenses or individual line accounts as a percent of net sales.

Operating Margin

> Formula: Operating Earnings divided by Net Sales
> Operating Earnings/Net Sales
> Operating Margin= $558,413 / $7,300,000 = 7.6%

This is viewed monthly and on a Year-to-Date basis. YTD is smoother versus a month's snapshot that may contain some odd items. But in all cases, explanations are important.

At this point, let me repeat the several levels of earnings as has been demonstrated in Chapter Four on Financial Statements; including the definition of operating earnings. Again, it is that level of earnings after the cost of sales has been subtracted (to reach the gross profit level); and after also subtracting those direct and indirect expenses associated with running the business such as selling expenses, shipping and receiving, general and administrative expenses, and other groups of expenses.

In this situation, it is not normal to annualize the ratio. However, it should be noted that if a business is highly seasonal, fixed expenses can have a significant impact on this level of expenses; reducing or deteriorating the operating ratio during a slow period as the fixed expenses will be a higher proportion of the costs than in periods of above monthly average sales. All income statement measurements key in on business activity (sales) and are either compared to it or are a result of it.

Further note that below the operating earnings level are generally "Other Income and Deductions." When this is subtracted, which is generally the case, from the operating Income line, pre-tax earnings results. From this, income taxes are subtracted resulting in Net Income. This is the situation in "C" corporations; not "sub-chapter S" corporations, which are taxed differently. Other forms of ownership will cause different situations for income tax calculation and collection. However, a reasonable tax rate, depending upon the form of legal organization, should be applied to generate the Net Income. This will serve for other calculations, especially "Return on Equity" (Net Income is divided by the Net Worth) and the Net Profit Margin percent.

Net Profit Margin = Net Income (after taxes) / Net sales
Net Profit Margin = $302,843 / $7,300,000 = 4.1%

The next set of ratios combine the income statement with the balance sheet.

Return on Total Assets

The Formula: Operating Earnings divided by Total Assets. The relationship of earnings produced from utilizing or investing all the company's assets including current assets, machinery, equipment, buildings, vehicles, computer equipment, and other assets will answer the following question. Is the company enjoying a return on its assets employed in the business operation?

Return on Assets = Operating earnings / Total assets
Return on Assets = $558,413 / $2,322,922 = 24.0%

Also in the analysis, partial year analysis should be annualized (in the same manner as was calculated to annualize Inventory Turnover) to produce a meaningful ratio or utilizing a rolling twelve months average of operating earnings by averaging 12 months of Total Assets.

Return on Equity (ROE)

This is also key.

The Formula: Net income (after taxes) is divided by the net worth of the company. This is the relationship of the return on the owner's own investment.

$$\text{ROE} = \text{Net Income (after taxes)} / \text{Net Worth (equity of the company)}$$
$$\text{ROE} = \$302{,}843 / \$979{,}213 = 30.9\%$$

The Net Income figure is the result of all the efforts of the company, or what is returned to the owners for their investment. If someone were to put money into a savings account and receive 4%, this is the comparable analysis. For example, in the example above, the net earnings of the company are $302,843, the Net Worth of the company at the beginning of the year was $979,213, and the Return on equity (ROE) is 30.9%.

The part of the ROE that may not be immediately seen (to the ownership) is in the market value of the company. Thus, if the ROE on operations is "satisfactory," what is not apparent is the additional benefit of the company's sale at a premium over the Net Equity balance on the books. This could be included in the market price of the stock, if publicly traded or not. Most likely, it is not until an offer is made for the company, then the stock price will move to reflect the offer. See the chapter on "Valuing the Company", Chapter Eighteen.

Finance Ratios

Debt Ratio

The Formula: Total Debt divided by Total Assets. This outlines how much of your assets are financed by other than equity. It is also used to convert the returns received to those received for ones own investment.

$$\text{Debt Ratio} = \frac{\text{Total Debt}}{\text{Total Assets}}$$

$$\text{Debt ratio} = \$1,343,708 / \$2,322,922 = .578$$

Times Interest Earned

The Formula: Operating earnings divided by interest expense. This is to reveal how many times the interest (a very fixed expense that must be met) is covered by operating earnings; i.e. earnings from running the business for which the financing and interest were contracted. Financing institutions are most interested in this number—as well as the owner(s). Are the interest disbursements able to be met and how many times over?

$$\text{Times Interest Earned} = \frac{\text{Operating Earnings}}{\text{Interest Expense}}$$

$$\text{Times Interest Earned} = (\$558,413 - \$36,500) / \$56,000 = 9.32$$

(Please note: this is a more conservative calculation as the bad debt amount is subtracted from the Operating Earnings.)

It may be that you will choose only a few of these formulas for your business or department. Many may not even relate to your business. Others will become valuable and you will key on those; either on a regular basis or just periodically check their trend.

They may be used by themselves or in comparison to other companies within your industry. You may simply want to compare them to what was developed within your financial plan (see Chapter Ten) or compare them to prior periods to analyze your company's trends. Principally, it would be used to spot why your returns are not as good as you had envisioned and areas to be improved. Remember, the return on equity can always be measured against a Treasury Bill of 30 years, a U.S. Government security, which is considered one of the "safest investments" and thus, is one of the lowest returns on an investment.

CHAPTER NINE —
HISTORICAL REVIEW

First Steps in Planning

Historical Review is a technique of comparison used to review past performance. It enables the reader to "line up" past years Income Statements (adjusting the line locations and definition to the latest statement) to view the flow of the dollars spent by account, and group of accounts, and the percent of sales each represents.

The goal is to analyze why each account changed, or didn't change, with the business activity over the period analyzed. If the line definitions remained the same over the period, management could better identify what accounts are variable versus those that do not change with volume. If account definitions changed during the period, then an attempt to adjust the past to the current basis should be made. Sometimes in these situations, estimates have to be made. The review will also identify those accounts with unusual activity occurring within them; noting and adjusting the "unusual activity."

The technique is rather simple. Line up the last three full years of a company's income statement, both dollars and percentages, and look across the line. The first key is to examine the percentage of sales for each line item. Dollars would be next.

Many accounts should vary directly with sales: direct materials, operating supplies, direct labor, payroll taxes, sales commissions, and similar items. Those similar items will become apparent as you view each account. With some accounts, detailed and nitty-gritty accuracy may be important. With other accounts, a broader, less detailed view should suffice.

The initial review and concentration should be on the elements of Cost of Sales to Net Sales (percentages/ratios). Movement of the percentages can reflect a number of important items.

1. **Direct material (as a percent of net sales):** Changes can reflect selling price changes. A reduction in the percent would indicate increased selling prices, or the following:

   - An increase in the percent may indicate that prices are not keeping pace with increases in material costs.
   - A material substitution.
   - Efficiencies may have changed, for the better or worse.
   - There may have been a change in the assortment of products sold, including the introduction of new products or the elimination of old products at different margins.
   - Customers may achieve volume breakpoints, if there are such programs.

   A sales listing/analysis will detail an assortment change. The assortment change could impact the margins because each product has different gross profit margins, or because of any other changes in the products sold (except for the selling price).

   Selling price changes can be determined by: (a) dividing the product sales by the units shipped; or (b) dividing the total cost of the product line by the unit product cost to determine the units and then go back to "(a)." Announced changes and the result of what actually took place in the billing system can be similarly analyzed. Price decreases may also result from product maturity or new competition. We will discuss product cost and profitability in Chapter Fourteen.

2. **Direct labor (as a percent of net sales):** This is also a barometer to be scrutinized. Direct labor can advise a change in staffing which occurred or didn't occur. What did happen? Some of the possibilities of why direct labor percent (of net sales) would have changed include the following:

   - Breakdown of equipment

- Introduction of new more efficient equipment
- Startup of same
- Product mix (assortment; the difference in labor required per product vs. the product selling price)
- Simple inefficiency
- Employee turnover
- Labor rate changes
- Inefficiencies resulting from raw material quality problems

3. **Selling and General Administrative Expenses:** Another area to be scrutinized, includes the following:

   - Commissions: This could result from policy changes or quota breakpoint changes because otherwise, the percent of sales should remain the same.
   - Advertising: While this should be examined on a percent of sales basis, the dollars spent may also be impacted by the programs started, when the results were felt, or simply by a change in the policy of the company.
   - Personnel: Customer service representatives may change on a step function basis (i.e., it is unreasonable to expect that time or personnel will move smoothly with sales). With small businesses, it is more likely that personnel will be restricted by the ability for a customer service representative to handle so many sales dollars; then becoming "inefficient" with the addition of personnel. Part time may be an interim answer, but it's still a step function. As sales grow, a reduced percent of sales will again result.

While these are just a few of the potential results derived from this historical comparison, I would hope you now see the potential from this analysis. It is an eye-opener. The what and why questions are detailed and explained regarding past operations. It should also serve as the basis for future projections; although, this would not be my preference in preparing a sales forecast.

No further analysis will be made here. It is up to you, the reader to take this technique to conclusion. Please review the example presented below.

Harvey Goldstein

XYZ Company Income Statement Review						
	Year-3		Year-2		Year-1	
Net Sales	4,441,925	100.0%	4,812,464	100.0%	6,679,709	100.0%
Cost of Sales						
Direct Materials	3,238,167	72.9%	3,329,439	69.2%	4,669,772	69.9%
Direct Labor	665,321	15.0%	632,055	13.1%	827,992	12.4%
Plant Expenses						
Utilities	38,184	0.9%	40,857	0.8%	50,254	0.8%
Depreciation	21,144	0.5%	23,544	0.5%	25,944	0.4%
Rent	35,784	0.8%	37,573	0.8%	39,452	0.6%
Health Ins.	94,500	2.1%	99,225	2.1%	102,202	1.5%
Shop Supplies	15,096	0.3%	15,398	0.3%	20,479	0.3%
Payroll Taxes	97,802	2.2%	92,912	1.9%	121,715	1.8%
Total Plant Exp	302,510	6.8%	309,509	6.4%	360,046	5.4%
Total Cost of Sales	4,229,998	95.2%	4,271,003	88.7%	5,857,810	87.7%
Gross Profit	211,927	4.8%	541,461	11.3%	821,899	12.3%
General & Administrative Expenses						
Officers Salaries	205,438	4.6%	220,535	4.6%	218,330	3.3%
Payroll Taxes	18,263	0.4%	21,437	0.4%	24,503	0.4%
Commissions	41,752	0.9%	53,724	1.1%	80,444	1.2%
Advertising & Promotion	7,905	0.2%	18,978	0.4%	25,611	0.4%
Auto Expenses	17,137	0.4%	17,775	0.4%	16,819	0.3%
Misc Taxes & Licenses	6,359	0.1%	3,920	0.1%	4,235	0.1%
Office Expense/Postage	16,841	0.4%	19,999	0.4%	22,978	0.3%
General Insurance	23,803	0.5%	28,338	0.6%	27,655	0.4%
Repairs & Maintenance	1,323	0.0%	2,220	0.0%	976	0.0%
Dues & Subscriptions	981	0.0%	1,495	0.0%	2,280	0.0%
Travel, Meals, & Ent.	22,450	0.5%	28,968	0.6%	35,843	0.5%
Telephone & Utilities	18,012	0.4%	13,109	0.3%	15,500	0.2%
Miscellaneous	26	0.0%	52	0.0%	41,000	0.6%
Total General & Admin.	380,290	8.6%	430,550	8.9%	516,174	7.7%
Income from Operations	(168,363)	-3.8%	110,911	2.3%	305,725	4.6%

Account Movement Characteristics: Fixed, Variable, and Partially

Variable Accounts

In the last chapter, the historical review provided insight regarding past performance, including operating expenses and how they relate to sales. The next step will be to identify all accounts as to their flexibility. The accounts will be identified as being either variable, fixed, or partially variable. This will greatly impact planning and understanding regarding the business' financial performance and controls that should be in place. An explanation of account types follows below:

Variable accounts: The amount spent will change with the sales and business activity. In most cases, the relationship is direct (i.e., a percent change in sales will cause this account to experience the same percent change in spending). The change experience may be "immediate," as with material usage; increased sales will result in increased usage, or reduced sales will result in reduced usage. An account may also be variable, but the expense may be delayed for a short period of time depending upon the account (e.g., supplies or commissions). However, aside from the timing, these accounts will move with sales and business activity.

Fixed accounts: A fixed account does not experience any direct changes as a result of an increase in sales (i.e., rent, administrative salaries, electrical power, water and garbage, etc.). However, in the long run, fixed accounts are also variable. For example, if sales were to grow significantly the current facility could become inadequate causing an increase in rent for an expanded facility or a replacement facility. These accounts will change on a "step basis," that is, not smoothly with business performance, but with significant increase when a major event occurs, and then remain flat or fixed for some period thereafter.

Partially-variable accounts: Partially-variable accounts usually have two components within them. One part is fixed and the other is variable. Telephone expense is a good example. The fixed portion usually reflects the line charges while the variable portion reflects the call volume, which is associated with business volume. In most cases, the split is about 50/50. However, when the split is dynamically different, examine the source of the charges to determine what the split should be. A "new company" must exercise judgment relative to the characteristics of each new account.

Note: Balance sheet accounts can also be defined accordingly. Accounts receivable and accounts payable will also vary with sales. Productive equipment (fixed assets) will not. Again, remember that no account is pure or fixed in the long run.

Budgeting and Financial Planning: May be one of the most important functions discussed. It provides the means to see the whole picture and see it "in advance." It sets the expectations, asks the question, "What if?" And responds, provides control, and the ability to adjust. It also provides a powerful communications medium for the organization. When managers are involved, it enhances their feeling of ownership; it enables them to contribute their knowledge, and provides for their own self-measurement. AND, how will you know you are getting to where you are going; unless, you know where you are supposed to go?

There are many methods to use for budgeting and planning. Two methods are discussed. One can be described as fast and broad in scope, while the other is much more detailed and defined. The latter method is preferred.

Change in Sales: The fast and broad-based method is called Change-in-Sales Method. In this method, little attention is paid to detail accounts and their change. The goal is to quickly forecast the sales and the fallout from it. Sales forecasts should not reflect any dynamic growth or the whole exercise will not be applicable.

The first item to estimate, in all situations, is the year-end estimate for the current year. However, this can be avoided if all one wants to see is what may occur in the forecast period. When forecasting for a particular time period assume that all accounts (in the Income Statement) are variable. Thus, if sales increase 10%, so should the bottom-line. For example, the percent earnings will remain the same, but the earnings dollars will change with the sales change. If sales are increasing, this represents a conservative estimate as percent net profit improvement should be experienced in reality. This happens because the fixed costs will remain unchanged with a growth in sales (leverage). Thus, the earnings percent and dollars should be higher.

However, if sales decline, the opposite would occur. Sales decline and the bottom line declines proportionally. In reality, fixed costs will not decline, the reduced bottom line will be worse than a proportional reduction. A decline is an unlikely situation for this method of forecast. If such an estimate is being considered, it is very important and the detail method would be used to plan out the consequences and how the company will operate to protect itself. Following is such a situation for your examination:

Sales have increased 15% over the current rate at the point of the year we are now in. **Seasonalize**, and project the Sales and Income Statement data to derive a year-end figure. If there is no seasonality, then a simple year-end projection should be sufficient, using current net profitability. (With a seasonal business it is beneficial to be more precise with year-end bottom line percentages.)

In Chapter Eight, <u>Financial Ratios</u>, certain mid-year data is projected to become annualized data. Below is a similar discussion regarding seasonalizing mid-year data to project year-end data. The methodology is the same, and is reiterated here.

Note: To project a seasonalized number to year end, use the following simple calculation. Divide the amount that is to be seasonalized by the percent of the year completed, by the month. For a straight, non-seasonalized number use the number of months completed to twelve. For example, in a calendar year, a May YTD number were to be seasonalized, then the factor would be 5/12 = .417. Thus, amount or number is 2,000,000 at the end of May (completed five months); then, 2,000,000/.417 would project to 4,796,163 for the whole year. With many businesses, there is some seasonality. For example, your business may be heavier from May through October or very heavy for the Christmas season, November and December, etc. In order to develop good seasonality numbers, a three-year experience chart should be developed. If that is not possible, then trade association statistics or similar data should be developed. See Chapter Thirteen for the development of seasonality data.

Harvey Goldstein

If the seasonalized experience, May YTD percent of total year is 38.8%; versus the 41.7% cited above, then the total year projection would change. Thus, the 2,000,000 number divided by .388 projects to 5,154,639, versus the straight-line projection of 4,796,163.

Continuing with the Change-in-Sales forecast. First, what is our goal and the goal of the exercise? One answer is simply, "I want to know where we will be, quickly, with a sales increase of 15%? How much will I make, what will my cash/financing position be?"

What would happen if:
1. Sales increase by 15%? The premise is that there are no changes in market or operations.
2. Net profit (after taxes) increases by 15%? A supporting premise that is the result of sales growth.
3. Accounts receivable increases by 15%? There would be no change in collection experience or DSO.
4. Inventory also moves with sales? This is a most likely scenario and is quick and dirty.
5. Other receivables will probably not move with sales.
6. Accounts payable increases by 15%? There would be no change in disbursement policy.
    a. Other current liabilities have to be reviewed, regarding whether they will move with sales. For example: accrued payables and taxes payable. However, the current portion of long term debt will not; as well as, loans to employees and officers, etc. will also not change with sales.

    b. Your revolving credit line should not be considered. Actually, this is partially a result of this quickie forecast, as its goal is to determine whether cash is generated or required for the planning period.

    c. Thereafter, fixed assets and long-term debt are not likely to move with a change in sales, unless the assets or equipment are at capacity. This could be a limiting production and sales factor or one requiring an infusion of cash to achieve the estimated desired growth.

So, what do we have? We have a sales forecast, an earnings forecast, cash drains due to increases in receivables and an inventory offset by increases in certain payables. We, most likely, do not have any movement in other accounts except for specific equipment investment that we may know about (rolling stock or autos and other required assets are sometimes considered "must spending"). We also have to add back the depreciation and amortization (non-cash charges) included within the Income Statement, which also is a favorable cash item. This was discussed in Chapter Five, Depreciation and Taxes.

After all is summed up, we have an excess of cash or a shortfall of cash. Dividends should be the last item addressed. Thus, we now have an operations forecast and a cash forecast for the forthcoming year on an annual basis. So, will we have to secure funding to be able to operate beyond our current capacity, pay dividends, or some combination? This will depend upon the company's policy.

The example below reflects what the above discussion could look like!

Cash Impact is Favorable or (Unfavorable) XYZ Company Change in Sales Forecast				
	01-Aug-xx	Total 2xxx	2xxy	Cash Impact
Income Statement Forecast				(unfav.)
		Growth:	15%	
Sales	1,200,000	1,800,000	2,070,000	
Net Earnings	48,000	72,000	82,800	
Depreciation/Amortiz.	30,000	45,000	51,750	
Cash from Operations	78,000	117,000	134,550	134,550
Balance Sheet Forecast				
Current Assets				
Cash				
Accounts Receivable	225,000	225,000	258,750	(33,750)
Inventories	257,000	257,000	295,550	(38,550)
Prepaid Expenses	20,000	0	0	0
Total Current Assets	502,000	482,000	554,300	(72,300)
P, P & E; orig. cost	225,000	225,000	225,000	
Pit, Prop & Equip	150,000	135,000	83,250	
net of Accumulated Depreciation				
Total Assets	652,000	617,000	637,550	
Current Liabilities				
Accounts Payable	100,000	100,000	115,000	15,000
Accrued Liabilities	15,000	15,000	17,250	2,250
Short Term Bank Loan	40,000	40,000	40,000	0
Other Payables	5,000	5,000	5,750	750
Total Current Liabilities	160,000	160,000	178,000	18,000
Notes Payable	50,000	40,000	10,000	(30,000)
Total Liabilities	210,000	200,000	188,000	(12,000)
			Cash Flow	50,250

In the above example, the result is that this company generated cash of $50,250. A good year. However, if the forecast determined a negative cash flow, additional funds would be required (or costs eliminated or sales increased, etc.). These funds could be secured via funds from the current ownership, sales of additional ownership shares, or additional borrowing. Further, no dividends were forecasted in this example. However, if the owners require it, even more funds will be required.

Please note that this was performed for a year in total. This cash forecast does not detail the monthly swings that can be severe, and could cause significant shifts in cash position. In most cases, one person can perform the forecast with a minimum of time and effort. But, "you get what you pay for."

Detailed Financial Plan (Preferred)

The Detail Forecast Method is what most companies utilize. It also starts with a sales forecast. However, in this instance, greater thought is put into the forecast. It may involve forecasts by product group or product line, the maturity of the products involved, new product inroads, targeting accounts, and any number of other means of creating the best forecast possible. It could also be influenced by external economic forces, the general economy, and other factors impacting the industry and your company. It may also include pricing assumptions. This is then reviewed (and usually approved by senior management) and distributed to the disciplines necessary including:

- Production for material and supplies requirements and labor requirements;
- Purchasing for pricing and material vendors and quality;
- Customer Service for staffing requirements; and
- Engineering for equipment purchase, maintenance and upgrade.

Budgets for the Sales Department and General Administrative Department(s) have to be developed along with certain policies regarding commissions, benefits, etc. And, most likely, other items and categories not noted here. Thus, a detailed line-by-line budget and financial plan is developed, usually taking two to four months to complete, as it is developed during regular operations and due to the sequencing of information that is acquired. A shorter time frame can be managed in a crunch situation. The detail plan also becomes the financial section of an overall business plan.

As part of the financial plan development, the Three-Year Historical Review may also be distributed to the various managers to provide a basis for new budgets. Purchasing and inventory plans are coordinated to meet the sales forecast, to insure the proper distribution within the inventory, and to account for lead time for any other considerations. A collections and disbursements policy review is also performed. New equipment and space requirements are reviewed by the appropriate personnel. Cost reductions, both achieved and proposed (conservative, even with proven equipment) are incorporated. Monthly sales seasonality is also reviewed as this also impacts the planning. Some companies project a percentage growth (usually based upon some premise). The monthly percentage growth is based upon shipments per day, times the number of shipping days in a particular month.

Budgets can be based upon experience, be "zero-based", or the amount can be dictated based upon someone's whim or experience. It is extremely important to ensure that each account is clearly defined. It might be beneficial to literally define what is in each account: all asset accounts, liability accounts, equity, sales, costs, and expenses. In all cases, the data is justified, but not yet approved. The budget is also based upon the most likely forecast that, generally, comprises the initial approved forecast.

The Finance Department, aside from managing the process and contributing towards certain budget items, must also generate the depreciation data and other plan data as required. We now have a detailed annual plan with solid bases for each line.

Based upon the seasonality of the monthly forecasts and the variability of each account, monthly plans are developed. Budget line items can vary with sales as with "cost of sales," can be set to be equal each month, or can be timed to be incurred in the proper month as with property taxes, bonuses or even a Christmas party.

At this point, we should have a financial plan for each month from sales through earnings. If this has been put on a spreadsheet computer program, with the relationships formulated, we can vary the sales (costs, expenses, price assumptions, etc, even the seasonality) and be able to play "what if" within reasonable deviations; and quickly get the results. If the "what if" variations become too wide (20% or more), request the managers for additional input regarding the impact on their discipline.

The balance sheet items will also be planned accordingly. For example, will the collections policy remain the same? Will new customers be sold under new terms, or will the company as a whole change the terms to all its customers? Will the disbursements policy change by taking the net days, versus stretching the payables, or taking discounts? What loans will be paid, or a new one negotiated or leases negotiated? What tax deposits will be made and when will they be made?

Once these forecasts and policies have been developed, the earnings can be brought all the way to cash: generated and/or required, by month. Peaks and valleys will be detailed. The managers have all had their input and can be held responsible for their commitment. At this point, the protective cushions must be viewed.

It must be noted that there are peculiarities within every industry and every business. However the industry or business operates, a plan can be developed and formulated; preferably on a computer, which is the best method.

Presented below is the financial plan for a manufacturer. It is an excellent source to base your own management questions, controls, and planning. While examining it, consider the following:

1. The plan is based upon the Three-Year Historical Review, outlined in a preceding chapter.
2. The seasonality discussed above (and detailed in Chapter Thirteen, Financial Tools) is also incorporated.
3. It projects the cash flow and defines the cumulative cash-flow line as the checkbook balance. In this example, the monthly cash-flow pays down the Line of Credit and the cash balance on the Balance Sheet remains the same.
4. No "f' or "v" for fixed or variable is noted; however, that should be easily identified.

Note: what is presented below is a matter of style. There are as many variations as there are preparers, but the principals are the same.

Three statements are presented:

XYZ Company
2xxy Financial Plan

sesasonalty	Actual 2xxx		Tot2xxy	%Sales	January	February	March	April	May	June	July	August	September	October	November	December	Total
					6.9%	7.6%	8.5%	8.3%	7.5%	8.1%	9.6%	8.2%	9.0%	9.0%	9.0%	8.2%	100.0%
Net Sales	6,679,709	100.0%	7,300,000	100.0%	505,973	554,057	622,904	607,605	545,314	591,213	703,772	601,048	655,689	655,689	655,689	601,048	7,300,000
Cost Of Sales																	
Direct Materials	4,669,772	69.9%	4,972,415	68.1%	344,645	377,397	424,293	413,872	371,442	402,706	479,377	409,405	446,624	446,624	446,624	409,405	4,972,415
Direct Labor	827,992	12.4%	865,252	11.9%	59,972	65,671	73,831	72,018	64,635	70,075	68,416	71,241	77,717	77,717	77,717	71,241	865,252
Plant Expenses																	
Utilities	50,264	0.8%	55,279	0.8%	4,607	4,607	4,607	4,607	4,607	4,607	4,607	4,607	4,607	4,607	4,607	4,607	55,279
Depreciation	25,944	0.4%	26,227	0.4%	2,186	2,186	2,186	2,186	2,186	2,166	2,186	2,186	2,186	2,186	2,186	2,186	26,227
Rent	39,452	0.6%	41,424	0.6%	3,452	3,452	3,452	3,452	3,452	3,452	3,452	3,452	3,452	3,452	3,452	3,452	41,424
Health Insurance	102,202	1.5%	107,312	1.5%	8,943	8,943	8,943	8,943	8,943	8,943	8,943	8,943	8,943	8,943	8,943	8,943	107,312
Shop Supplies	20,479	0.3%	22,381	0.3%	1,551	1,699	1,910	1,863	1,672	1,813	2,158	1,843	2,010	2,010	2,010	1,843	22,381
Payroll Taxes	121,715	1.8%	127,192	1.7%	8,816	9,654	10,853	10,567	9,501	10,301	12,262	10,472	11,424	11,424	11,424	10,472	127,192
Total Plant Exp	360,046	5.4%	379,816	5.2%	29,554	30,539	31,950	31,636	30,360	31,301	33,607	31,502	32,622	32,622	32,622	31,502	379,816
Total Cost of Sales	5,857,810	87.7%	6,217,482	85.2%	434,171	473,608	530,074	517,526	466,437	504,082	596,400	512,148	556,963	556,963	556,963	512,148	6,217,482
Gross Profit	821,899	12.3%	1,082,518	14.8%	71,802	80,449	92,830	90,079	78,877	87,131	107,373	88,900	98,726	98,726	98,726	88,900	1,082,518
%sales					14.2%	14.5%	14.9%	14.8%	14.5%	14.7%	15.3%	14.8%	15.1%	15.1%	15.1%	14.8%	14.8%
General & Administrative Expenses																	
Salaries (Incl. Pit. Mgr)	218,330	3.3%	248,420	3.4%	20,702	20,702	20,702	20,702	20,702	20,702	20,702	20,702	20,702	20,702	20,702	20,702	248,420
Payroll Taxes	24,503	0.4%	26,951	0.4%	2,246	2,246	2,246	2,246	2,246	2,246	2,246	2,246	2,246	2,246	2,246	2,246	26,951
Commissions	80,444	1.2%	86,963	1.2%	6,028	6,600	7,420	7,238	6,496	7,043	8,384	7,160	7,811	7,811	7,811	7,160	86,963
Advertising & Promotion	25,611	0.4%	26,000	0.4%	10,000							8,000		8,000			26,000
Auto Expenses	16,819	0.3%	18,000	0.2%	1,500	1,500	1,500	1,500	1,500	1,500	1,500	1,500	1,500	1,500	1,500	1,500	18,000
Misc. Txs & Licenses	4,235	0.1%	3,500	0.0%				500	2,000							1,000	3,500
Office Expense/Postage	22,978	0.3%	24,000	0.3%	2.000	2,000	2,000	2.000	2,000	2,000	2,000	2,000	2,000	2,000	2,000	2,000	24,000
General Insurance	27,655	0.4%	32,250	0.4%	2,500	2,500	2,500	2,500	2,500	2,500		7,250	2,500	2,500	2,500	2,500	32,250
Repairs & Maintenance	976	0.0%	1,200	0.0%	100	100	100	100	100	100	100	100	100	100	100	100	1,200
Dues & Subscriptions	2,280	0.0%	2,400	0.0%		1,000	100	100				100		100	1,000		2,400
Travel, Meals & Entertainment	35,843	0.5%	36,000	0.5%	3,000	3,000	3,000	3,000	3,000	3,000	3,000	3,000	3,000	3,000	3,000	3,000	36,000
Telephone & Utilities	15,500	0.2%	16,021	0.2%	1,335	1,335	1,335	1,335	1,335	1,335	1,335	1,335	1,335	1,335	1,335	1,335	16,021
Miscellaneous	41,000	0.6%	2,400	0.0%	200	200	200	200	200	200	200	200	200	200	200	200	2,400
Total Gen & Admin	516,174	7.7%	524,105	7.2%	49,610	41,183	41,103	41,421	42,079	40,626	39,467	53,593	41,394	49,494	42,394	41,743	524,105
% sales					9.8%	7.4%	6.6%	6.8%	7.7%	6.9%	5.6%	8.9%	6.3%	7.5%	6.5%	6.9%	7.2%
Income from Operations	305,726	4.6%	558,413	7.6%	22.192	39,266	51,727	48,656	36,798	46,505	67,906	35,307	57,332	49,232	56,332	47,157	558,413
%sales					4.4%	7.1%	8.3%	8.0%	6.7%	7.9%	9.6%	5.9%	8.7%	7.5%	8.6%	7.8%	7.6%
Other Income & Deducts																	
Inte sl: Expense	(62,000)	-0.9%	(56,000)	-0.8%	(4,667)	(4,667)	(4,667)	(4,667)	(4,667)	(4,667)	(4,667)	(4,667)	(4,667)	(4,667)	(4,667)	(4,667)	(56,000)
Bad Debts	56,000	0.8%	36,500	0.5%	2,530	2,770	3,115	3,038	2,727	2,956	3,519	3,005	3,278	3,278	3,278	3,005	36,500
Total Other Income / Deducts	(118,000)	-1.8%	(92,500)	-1.3%	(7,197)	(7,437)	(7,781)	(7,705)	(7,393)	(7,623)	(8,186)	(7,672)	(7,945)	(7,945)	(7,945)	(7,672)	(92,500)
Pre-tax Income	187,726	2.8%	465,913	6.4%	14,996	31,829	43,946	40,953	29,406	38,883	59,721	27,835	49,387	41,287	48,387	39,485	465,913
Income Taxes 0.35	65,704	1.0%	163,069	2.2%	5,248	11,140	15,381	14,334	10,292	13,609	20,902	9,672	17,285	14,450	16,935	13,820	163,069
Net Income	122,021	1.8%	302,843	4.1%	9,747	20,689	28,565	26,620	19,113	25,274	38,818	17,963	32,101	26,836	31,451	25,665	302,843
%.sales					1.9%	3.7%	4.6%	4.4%	3.5%	4.3%	5.5%	30%	4.9%	4.1%	4.8%	4.3%	4.1%

XYZ Company
2xxy Financial Plan

	Actual 2xxx	Tot2xxx	%Sales	January	February	March	April	May	June	July	August	September	October	November	December	TOTAL
Cash Flow																
MEMO: Inventory Purchase Plan				361,021	400,845	419,082	392,657	387,074	441,041	444,391	428,015	448,024	446,824	428,015	394,257	4,989,647
Callections				500,457	527,245	526,900	585,443	612,528	573,504	564,745	644,487	649,132	625,090	652,410	652,683	7,114,623
Operating Disbursements																
Materials				303,191	361,021	400,845	419,082	392,657	387,074	441,041	444,391	428,015	446,624	446,624	428,015	4,898,581
Mfg Labor				59,972	65,671	73,831	72,018	64,635	70,075	83,416	71,241	77,717	77,717	77,717	71,241	865,252
Plan Exp; current month				25,817	26,655	27,855	27,588	26,503	27,302	29,264	27,474	28,426	28,426	28,426	27,474	331,208
Plant Exp 30 days				1,542	1,551	1,699	1,910	1,863	1,672	1,813	2,158	1,843	2,010	2,010	2,010	22,080
G & A Current				27,648	27,648	27,648	27,348	27,648	27,648	25,148	32,398	27,648	27,648	27,648	27,648	334,021
G & A 30 days				9,880	21,963	13,535	13,456	13,773	14,431	12,978	14,319	21,195	13,746	21,846	14,746	185,869
Interest Expense				4,667	4,667	4,667	4,667	4,667	4,667	4,667	4,667	4,667	4,667	4,667	4,667	56,000
Loan repayments				9,300	9,300	9,300	9,300	9,300	9,300	9,300	9,300	9,300	9,300	9,300	9,300	111,600
Capital Expenditures				7,000	7,000	2,000	2,000	2,000	2,000	2,000	2,000	2,000	2,000	2,000	2,000	34,000
Taxes Pay- 2000				5,650			5,650									
Tax Deposits							36,691		36,691			36,691				110,072
Cash Flow				45,790	1,770	(34,479)	(34,566)	69,483	(7,356)	(44,882)	36,541	11,631	12,952	32,172	65,584	154,641
Credit; paydown or draw up				(45,790)	(1,770)	34,479	34,586	(69,483)	7,356	44,882	(36,541)	(11,631)	(12,952)	(32,172)	(65,584)	(154,641)
Cumulative Cash Flow		9,650 Bai.Fr. 12/31/xx		9,650	9,650	9,650	9,650	9,650	9,650	9,650	9,650	9,650	9,650	9,650	9,650	

XYZ Company
2XXY Financial Plan

	Actual 2xxx	Total 2xxy	%Sales	January	February	March	April	May	June	July	August	September	October	November	December	TOTAL
Balance Sheet	as of 12/31/00															
Current Assets																
Cash	9,650			9,650	9,650	9,650	9,650	9,650	9,650	9,650	9,650	9,650	9,650	9,650	9,650	
Accounts Receivable	8,901,300			893,787	917,828	1,010,717	1,029,841	959,901	974,654	1,110,163	1,063,719	1,066,997	1,094,317	1,094,317	1,039,677	
Inventory	548,000			564,456	587,904	582,693	561,479	577,111	615,446	580,460	599,070	599,070	599,070	580,460	565,312	
Due from Officers	38,450			36,450	36,450	36,450	36,450	36,450	36,450	36,450	36,450	36,450	36,450	36,450	36,450	
Total Current Assets	1,464,980			1,504,343	1,561,832	1,639,510	1,637,420	1,583,112	1,636,100	1,736,723	1,708,188	1,712,167	1,739,487	1,720,878	1,651,089	
Plant, Property & Equipment	1,147,150			1,154,150	1,161,150	1,163,150	1,165,150	1,167,150	1,169,150	1,171,150	1,173,150	1,175,150	1,177,150	1,179,150	1,181,150	
Machinery & Equipment	75,610			56,610	75,610	75,610	75,610	75,610	75,610	75,610	75,610	75,610	75,610	75,610	75,610	
Office Furniture and Fixtures	1,222,760			1,129,760	1,236,760	1,238,760	1,240,760	1,242,760	1,244,760	1,246,760	1,248,760	1,250,760	1,252,760	1,254,760	1,256,760	
Total P, P &.E	558,700			560,886	563,071	565,257	567,442	569,628	571,614	573,999	576,185	578,371	580,556	582,742	584,927	
Less: Accumulated Deprec	664,060			668,874	673,689	673,503	673,318	673,132	672,946	672,761	672,675	672,390	672,204	672,018	671,833	
Total Assets	2,149,040			2,173,217	2,225,521	2,313,014	2,310,738	2,256,244	2,309,146	2,409,484	2,381,463	2,364,556	2,411,691	2,392,896	2,322,922	
Current Liabilities																
Curr port of Long term debt	88,520			88,520	88,520	88,020	88,520	88,520	88,520	88,520	88,520	88,520	88,520	88,520	88,520	
Accounts Payable - Trade	755,520			825,442	856,986	875,354	849,200	844,084	898,739	901,774	891,959	903,287	911,387	885,678	851,102	
Taxes Payable	11,300			10,896	22,039	37,420	9,413	19,704	(3 ,377)	17,525	27,197	7,792	22,242	39,178	52,998	
Line of Credit with Bank	425,600			379,810	378,040	412,519	447,085	377,602	384,958	429,839	393,298	381,667	368,715	336,542	270,959	
Total Current Liabilities	1,280,940			1,304,670	1,345,584	1,413,131	1,394,217	1,329,910	1,366,839	1,437,658	1,400,975	1,381,268	1,390,884	1,349,918	1,263,578	
Long Term Liabilities																
Notes Payable	191,730			182,430	173,130	163,830	154,530	145,230	135,930	126,630	117,330	108,030	98,730	89,430	80,130	
Total Liabilities	1,472,670			1,487,100	1,518,714	1,577,643	1,548,747	1,475,140	1,502,769	1,564,288	1,518,305	1,489,296	1,489,594	1,439,348	1,343,708	
Slockholders' Equity																
Capital Stock	437,480			437,480	437,480	437,480	437,480	437,480	437,480	437,480	437,480	437,480	437,480	437,480	437,480	
Retained Earnings	238,890			248,637	269,326	297,891	324,510	343,624	368,897	407,716	425,679	457,780	484,616	516,068	541,733	
Total Stockholders' Equity	676,370			686,117	706,806	735,371	761 ,990	781,104	806,377	845,196	863,159	895,260	922,096	953,548	979,213	
Total Liab & Equity	2,149,040			2,173,217	2,225,521	2,313,014	2,310,738	2,256,244	2,309,146	2,409,484	2,381,463	2,384,556	2,411,691	2,392,896	2,322,922	
Key Ratios / Data		Average 2xxy														
Net Working Capital	204,040			199,673	206,247	225,698	243,203	253,202	269,361	299,065	307,913	330,901	348,623	370,960	387,511	
Current Ratio	1.16			1.15	1.15	1.16	1.17	1.19	1.20	1.21	1.22	1.24	1.25	1.27	1.31	
Quick Ratio	0.73			0.72	0.72	0.75	0.77	0.76	0.75	0.80	0.79	0.81	0.82	0.84	0.86	
Times Fixed Charges	4.0		9.3	4.2	7.8	10.4	9.8	7.3	9.3	13.8	6.9	11.6	9.8	11.4	9.5	
Inventory Turns: YTD projected	10.69		10.64	11.10	10.85	10.78	10.85	10.85	10.72	10.71	10.67	10.63	10.60	10.61	10.64	
DSO			51.19	54.76	46.38	50.30	50.85	54.57	49.46	48.90	54.86	48.82	51.74	50.07	53.62	
Asset Turnover	10.1		10.9	10.9	10.9	10.9	10.9	10.9	10.9	10.9	10.9	10.9	10.9	10.9	10.9	
Debt Ratio	0.685		0.644	0.684	0.682	0.682	0.670	0.654	0.651	0.649	0.638	0.625	0.618	0.602	0.578	
Return on Equity	18.0%		36.6%	20.5%	29.7%	34.8%	35.8%	34.5%	34.3%	35.3%	33.4%	33.1%	32.2%	31.7%	30.9%	

1. An Income Statement: detailing the actual operations for the current year and the total forecast year. The Annual Plan is spread to months based upon the seasonality. Note the accounts that are fixed and variable. Also note that the bad debts (within the Other Income and Deductions section) and the income tax line that are in reality "provisions" for bad debts and income taxes. Actual disbursements are charged to the offsetting liability account on the Balance Sheet. All the line items (costs and expenses) offered are just a sampling of what would be included. The loan repayment is not amortized as to principal and interest in this example. Nor is the interest adjusted for the monthly balance outstanding for the Line of Credit.

2. A Cash Flow statement: it initiates with a memo item regarding inventory purchase receipts. In this example, the basis is purchases or receipts that are made for one-half the current months estimated usage and one-half the next month's usage. There are formulas or calculations for collections and disbursements. Here are a few examples:

   a. Collections are for one-half the last month and one-half the month prior less the bad debts provision included with in the Income Statement, "Other Deductions" section. This is approximately forty-six days.

   b. Materials received are paid for in net thirty days. Thus, what is received in May is paid for in June, approximately thirty days.

   c. All the salary, labor, and payroll taxes are current, (i. e., paid for in the month they are incurred).

   d. Plant and General and Administrative (G & A) expenses are either paid for in the month they are incurred or are paid for in thirty days (such as office supplies). Depreciation is not considered in this cash-flow statement, so there is no "add-back" necessary. It's the same, just a different approach for planning, not reporting. Estimated interest expense, per the Income Statement, has its own line.

   e. Loan principal repayments are detailed. (See comment regarding Interest calculation in the Income Statement, item above.)

    f. Capital expenditures are detailed, based upon the overall investment program. Tax disbursements are calculated upon 90% of one-fourth of the estimated liability. Payments are made in April, June, and September. The fourth payment or deposit is made in January of the following year, with the following April 15th tax filing experiencing the difference between the payment or deposits and the actual liability from the tax return.

This is then summed to the monthly and cumulative cash flow.

Note: In this example, the monthly cash flows will go to pay down or draw up the Line of Credit. Thus, the Balance Sheet Line-of-Credit Account is impacted accordingly, and the Cash Account balance remains constant throughout the period.

3. Balance Sheet: it initiates with the December 31, current year balances. As noted, the cash balance remains constant throughout the period.

- Accounts Receivable is increased by the monthly sales from the prior balance and is reduced by the collections.

- Inventory (increases upon the receipt of goods and decreases when shipments are made as casted on the income statement); accounts payable (increases upon receipt and decreases upon payment) and taxes payable are similarly impacted; taxes payable is reduced when the tax payment or deposit is made per the government schedule.

- Machinery and equipment increases by the capital expenditures, accumulated depreciation (a contra account) are increased by the depreciation expense.

- Notes payable is reduced by the payments in the cash flow forecast (no new borrowing is incorporated), retained earnings is impacted by the net earnings or loss.

- RATIOS (for the plan) are calculated for measurement, analysis, and to ensure that on a plan basis, the company is in good standing with any loan covenants.

These charts, statements, and explanations are offered to outline the mechanics and to reinforce the interaction of the accounting or finance interrelationships. Thus, when the actual statements are published, you will understand them much better. The above plan is a simple one, having relatively few accounts; but the numbers are down to the dollar to simulate realism, and to ensure numerical accuracy. Of course, when you check the math for yourself, you will enhance your understanding.

In summary, financial planning will lead to a better understanding of the business and the financial reports published. It also provides one key element in the financial controls of the company as it causes examination when variations occur.

Now that the "most likely" financial plan is approved, it will be input onto the main frame computer where the actual accounting is performed and all the financial records reside. Comparative financial statements will be produced.

CHAPTER ELEVEN —
COMPARATIVE ANALYSIS

Do you receive monthly financial statements? Does it detail the information that will let you know where you are and how well you are or are not doing? And if so, how would you know? The answer is to **compare and analyze**.

Comparisons are usually made against the budget or financial plan and against the same period in the prior year and on a year-to-date basis. Percentage analysis and comparison is one key element. This is the detailing of all Income Statement items against the Net Sales value. The Net Sales value is generally gross sales to customers (the billing amount of each invoice in total) less any returns (of product), discounts (from the billing price) and allowances (credits for various items). These are reductions or deductions from the gross sales and result in the net value received from the customers.

        GROSS SALES
        Less: Returns
                Discounts and allowances
        NET SALES

In all cases, in all situations, the initial analysis should be to examine the percentages. In all the Profit and Loss Statement presentations, percentages should accompany the hard dollar numbers.

For Profit and Loss Statement purposes, I would suggest something like the example at the end of this chapter. The example is a summary of the Income Statement and does not contain all the accounts within each grouping. Those may be reported on a separate schedule or incorporated within this. I chose not to incorporate it here. If the comparative statement cannot include the data, then the statement should have at least two of the data columns with percentages. You will notice that I have placed the description column in the center for ease of reading (in many cases, this column it usually found on the left).

As for the Balance Sheet, the comparison to plan and prior periods is also informative, including percent of sales. In many cases, the percent data is extraneous, e. g. cash as a percent of sales. However, in many circumstances, it is very informative, such as with accounts receivable, accounts payable, fixed assets, and total assets. These are the reciprocal of the formulas previously discussed and reflect efficiency, just on a different basis. These must be carefully interpreted especially regarding partial year data as discussed in the Chapter Eight on Ratios.

XYZCompany
Financial Statements as of
February 28, 2xxx

Income Statement

Description	Current Month: February 2xxx						Year-to-Date					
	Actual	% sales	Plan	% sales	Last Year	% sales	Actual	% sales	Plan	% sales	Last Year	% sales
Gross Sales	585,120	100.7%	557,000	100.5%	543,422	100.4%	1,007,837	100.7%	1,065,660	100.5%	968,079	100.6%
Less: Return	4,025	0.7%	1,835	0.3%	1,050	0.2%	6,175	0.6%	3,380	0.3%	3,552	0.4%
Disc. & Allow.	257	0.0%	1,108	0.2%	1,252	0.2%	804	0.1%	2,250	0.2%	2,407	0.3%
Net Sales	580,838	100.0%	554,057	100.0%	541,120	100.0%	1,000,858	100.0%	1,060,030	100.0%	962,120	100.0%
Cost of Goods Sold	488,021	84.0%	473,608	85.5%	463,740	85.7%	849,071	84.8%	907,779	85.6%	829,347	86.2%
Gross Profit	92,817	16.0%	80,449	14.5%	77,380	14.3%	151,787	15.2%	152,251	14.4%	132,773	13.8%
	85											
Sales & Marketing		0.0%		0.0%		0.0%		0.0%		0.0%		0.0%
Gen'l &Admin. Exp.	39,473	6.8%	41,183	7.4%	40,385	7.5%	84,630	8.5%	90,793	8.6%	80,585	8.4%
Operating Income	53,344	9:2%	39,266	7.1%	36,995	6.8%	67,157	6.7%	61,458	5.8%	52,188	5.4%
Other Inc. & Deduc.												
Interest Expense	(4,808)	-0.8%	(4,667)	-0.8%	(5,207)	-1.0%	(9,118)	-0.9%	(9,334}	-0.9%	(1,0,314)	-1.1%
Bad Debts	2,000	0.3%	2,770	0.5%	1,600	0.3%	3,000	0.3%	5,300	0.5%	2,900	0.3%
Tot. Oth. Inc & Ded	(6,808)	-1.2%	(7,437)	-1.3%	(6,807)	-1.3%	(12,118)	-1.2%	(14,634)	-1.4%	(13,214)	-1.4%
Pre-tax Earnings	46,536	8.0%	31,829	5.7%	30,188	5.6%	55,039	5.5%	46,824	4.4%	38,974	4.1%
Income Taxes	16;288	2.8%	11,140	2.0%	10,566	2.0%	19,264	1.9%	16,388	1.5%	13,641	1.4%
Net Earnings	30;248	5.2%	20,689	3.7%	19,622	3.6%	35,775	3.6%	30,436	2.9%	25,333	2.6%

Variance Reporting

Now that we have set up the initial comparative statement, let us move forward and analyze the results. This, together with the comparative report, is one of the keys to managing the business. "How did I perform?" versus, "How I planned to perform and why?" Also, "How am I doing versus last year, and why?"

The plan is set upon certain underlying economic premises. Therefore, if the performance is materially deviating for more than a two-month period, corrective action should be investigated. (One month is insufficient to use as a tripping devise. However, it is a warning devise). Interim reports, discussed in Chapter Thirteen are another means used as an early warning device.

Analysis can be made by simply taking the difference between the actual and the plan, for the month and for the year-to-date. The initial analysis would center upon the difference in sales on a dollar and percent of plan basis. Thus, the initial Variance Review may be: "Net Sales totaled $581,000 in the month, $27,000 (4.6%) below plan." Thereafter, the variation for each line is reviewed accordingly. The gross profit and totals for each expense category are also discussed when significant. Year-to-date figures are also reviewed. In some cases, they may be more revealing. The chart or Financial Statement (just presented) is the first step in the review process.

However, two additional views should be made. The first is an examination of the percent of sales (the % column) for each line item; Actual versus Plan. What is the planned percent of Net Sales for a particular line versus the actual percent? This will reveal the efficiency, particularly for the variable items that should move with the sales. The percents for the fixed expenses will move contra to the sales movement compared to the budgeted amount; for example, if sales are higher than planned and the fixed expense expenditures is as expected; then, as a percent of sales, the fixed expenses will be a smaller percent (a favorable movement of the percent and vice versa).

The second is to compare the planned fixed cost items versus the plan. This will ensure budget integrity for those line items. This is in addition to the percent of sales examination. Further, if sales are trending differently than was planned, future budgeted expenditures may have to be reviewed and adjusted, up or down.

For the variable, in theory, the percents should be as planned if there is only a volume difference. Reality says that in an increased volume situation; the labor is humming a bit better and the percent to sales should be a bit "favorable" to the plan (a lower percent to sales than the planned percent). In a reduced sales situation, the efficiencies will most likely deteriorate as "the job expands to fill the time allotted." Get the picture?

I used the term "favorable" rather than saying "under budget," as under budget expense is good and "under budget" sales are not. This is my preference.

There is **another analytical technique** that we can employ, "Budget Adjusted for Volume." In this technique, the "budget" column is set to the actual sales. Obviously the fixed items do not change, but the budgeted variable items change proportionately with the adjusted budgeted sales. Then, the percentage and dollar differences can be examined on a comparable basis. Thus, the efficiency differences are dollars as the plan data has been adjusted for the actual sales level.

There are many numbers of ways to complete the analysis, pick one and become comfortable with it. This budget adjusted for sales may be correct in theory; however, the impact on the reader of the analysis is not as great, as the actual sales are not detailed on the analysis.

Highly Summarized Report

At this point, we have comparative Financial Statements (with percentages) for the Income Statement and the Balance Sheet. We are also able to include certain financial ratios. In addition, we have analyzed the performance against the financial plan (the control) and against the prior year. All this data is presented in segments and in total to various managers and general management personnel. Depending upon the particular company and its style of management, a highly summarized report may also be produced, almost as a "cheat sheet." It generally contains five or so lines from the Income Statement (percentages and plan, month and YTD), and comparable key data from the Balance Sheet. This report also contains some bullets explaining several key operating information regarding sales, operations, expenses, collections, disbursements, etc. There should be no set pattern for the bullets. They should just best describe the month completed.

CHAPTER TWELVE —
CASH FLOW

While all financial data is necessary and critical, a company runs on cash. So, in this chapter we will look at cash flow with three distinct goals.

1. What has it been? The actual during the period.
2. What will it be during a planning period? An important part of the financial plan.
3. A reforecast, in detail, for the four to eight weeks just ahead. This involves creating a new special report.

Actual

The **Cash Flow Statement** is the third financial statement to be generated. Along with the Income Statement and Balance Sheet, it combines certain aspects of both. In general, most Cash Flow Statements have three sections:

- Operations
- Investing
- Financing

Operations

Operations encompass the Earnings Statement cash plus the change in those current assets and current liabilities associated with the daily operating of the business. This is because the Earnings Statement does not account for delays (or advances) in receipts or payments. Initially, Earnings Statement cash is the net earnings (after taxes) plus depreciation and amortization. These are non-cash expenses that are charged against earnings.

Net Earnings (after taxes)
<u>+ depreciation/amortization</u>

Cash Net Earnings

For example, unless a sale results in an immediate **receipt of cash** (as in a supermarket), the sale is recorded as valid, but no actual cash is received. It is accounted for as an account receivable. Thus, from a cash viewpoint, the Earnings Statement must be adjusted for collections of outstanding receivables. The difference in the balance of the receivables from the beginning of the period under review to the end of the period is incorporated into the Operating Activities section of the Cash Flow Statement.

The change in certain current assets and liabilities must be considered, since the Earnings Statement does not account for delays (or advances) in receipts or payments. In an extreme example, if a company's only activity was to sell one item for $100.00 on credit (no other expenses of any kind occurred in the month); then, the Earnings Statement would show a $100.00 profit (taxes are not considered). However, there is no actual cash received by the company! Unless a sale results in an immediate receipt of cash (as in a supermarket), the sale is recorded as valid, but no actual cash is received. It is accounted for as an account receivable. Thus, from a cash viewpoint, the Earnings Statement has to be adjusted for collections of outstanding receivables. In this case, a $100.00 profit is offset (from a cash standpoint) by the growth (unfavorable) of the accounts receivable balance by $100.00. Reflecting that no cash was received within the Operating Activities section of the Cash Flow Statement.

The accounts most associated with the Cash Flow from operating activities are accounts receivable, inventory, accounts payable, accrued payables, income tax liabilities and payroll tax liabilities and employee advances (there may be other accounts, but for that a finance professional should get involved). It should be noted that the current items associated with a line of credit or the current portion of any long-term debt are "financing" items to be handled in that section.

The cash flow calculation involves taking the difference between the balances, from the start of the period to the end of the period being reported (for the balance sheet accounts). Here is the Income Statement and Balance Sheet for the period from January 1 through March 31, 2xxx.

The ABC Company Cash Flow from Operating Activities Example January 1, 2xxx - March 31, 2xxx		
Income Statement		
Net Sales		$ 521,000
Cost of Goods Sold		
Materials		270,000
Labor		100,000
Plant expense		50,000
Depreciation		16,000
Total Cost of Good sold		436,000
Gross Profit		55,000
Selling, General & Administrative		51,667
Operating Earnings		33,333
Miscellaneous Income and Deductions		5,000
Pre-tax Earnings		28,333
Income Taxes - paid		11,333
Net Income		$17,000

The ABC Company		
Balance Sheet	As of 1/1	As of 3/31
Assets		
Cash	$1,500	$15,000
Accounts Receivable	260,000	293,000
Inventory	400,000	425,000
Loans to Employees	15 000	13 000
Total Current Assets	676,500	746,000
Long-Term Assets		
Automotive Equipment	75,000	85,000
Computer Equipment	125,000	130,000
less: total accumulated depreciation	(102,000)	(118,000)
Net Long-Term Assets	98,000	97,000
Total Assets	$774,500	$843,000
Liabilities		
Accounts Payable	$125,000	$150,000
Bank Line of Credit	250,000	271,500
Total Current Liabilities	375,000	421,500
Long-Term Assets		
Term Loan	125,000	110,000
Capital Lease on New Truck	0	20,000
Total Long-Term Assets	125,000	130,000
Total Liabilities	$500,000	$551,500
Stockholders Equity	$274,500	$291,500
Total Liabilities and Equity	$774,500	$843,000

BALANCE SHEET ACCOUNT ANALYSIS FOR CASH FLOW			
			Change
	January 1	March 31	Cash Inflow or (Cash Outflow)
			Favorable or (Unfavorable)
Accounts Receivable	$ 260,000	$ 293,000	($ 33,000)

Analysis: This is an **Unfavorable** movement

This is an outflow of funds as your company does not have the cash. You have done business, more business, great! But, have not yet collected for all the sales. More of your money is owed to you for whatever reason, in this case by the amount of $33,000. Yes, the terms of sale have not changed nor the average DSO, which means that the cash that should be in your pocket from the increased business activity is not, not yet. It must come from somewhere or disbursements are denied or delayed somewhere else. Compare this to a pure cash business, like a supermarket. Even when the sales are increasing, the cash would be collected at the point of sale.

	January 1	March 31	Change
Inventory	$400,000	$425,000	($25,000)

Analysis: This is an **Unfavorable** movement

This is also an outflow as there is more money (cash) invested in the inventory.

	January 1	March 31	Change
Accounts Payable	$125,000	$150,000	$25,000

Analysis: This is a *Favorable* movement.

This is an inflow because we have more of someone else's cash then we had before (as we have not yet paid our bills).

During the period, the employee's loan balance declined by $2,000 via repayment to the company of the loan. This is **Favorable**.

TOTAL   ($31,000)

Thus, from these Balance Sheet accounts (example above), we have an unfavorable movement (an outflow) of ($31,000). If, net earnings were $17,000 in the first three months, by adding back depreciation of $16,000 the total operations cash flow was an inflow of $2,000. See the total Cash Flow Statement below.

Investing

Investing encompasses the purchases of all assets for the business to operate. It also includes the divestment of same, for example, the purchase of trucks, computers, manufacturing equipment, sale of trucks, obsolete equipment, etc.

At this point, the reduction of the original (book) purchase price to the "net book value" reducing the capital equipment original cost accounts by the accumulated depreciation accounts should be avoided. Depreciation has been included as part of the cash flow from Operating Activities.

	January 1	March 31	Change
Automotive Equipment	$75,000	$85,000	($10,000)
Computer Equipment	$125,000	$130,000	($ 5,000)
Total			($15,000)

This is an **Unfavorable** movement. The company has expended the money.

Financing

Financing is the area of the report where the borrowing and repayments are detailed. Some questions to be addressed via the financing cash flow section are:

1. Were your purchases of equipment financed by a "capital lease"? (We'll discuss specifically Chapter Sixteen.)

2. Were there any new loans taken from an institution publicly or privately or were the loans reduced via payments?

3. Were dividends paid?

4. What happened to the revolving line of credit? (If this is an active line of credit, my preference is to show it as a separate line item.) Short term cash items (such as, marketable securities), may be included in the actual cash balances to sum to the "checkbook balance" as of a certain date, if it can be quickly converted into cash (such as, time deposits).

	January 1	March 31	Change
Capital lease for new truck	$ 0	$20,000	$20,000
Term Loan...Our Principal Bank	$125,000	$110,000	($15,000)
Line of Credit...Principal Bank	$250,000	$271,500	$21,500
Total Financing			$26,500

Ending cash balance ("checkbook balance")-IT MUST BE THIS BY ITS OWN DEFINITION, i.e. opening balance plus change equals current balance to the penny $15,000

Harvey Goldstein

Below is a "typical" Cash Flow Statement. Again, this is the third statement and is critical to management.

<table>
<tr><td colspan="2" align="center">ABC Company
Cash Flow Statement
For the Period Ending March 31, 2xxx</td></tr>
<tr><td></td><td></td></tr>
<tr><td>Cash Flow from Operating Activities</td><td></td></tr>
<tr><td>Net Income</td><td>$17,000</td></tr>
<tr><td>Depreciation</td><td>16,000</td></tr>
<tr><td></td><td></td></tr>
<tr><td>Change in Accounts Receivable</td><td>(33,000)</td></tr>
<tr><td>Change in Accounts Payable</td><td>25,000</td></tr>
<tr><td>Change in Inventory</td><td>(25,000)</td></tr>
<tr><td>Change in short-term (employee) loans</td><td>2,000</td></tr>
<tr><td></td><td></td></tr>
<tr><td>Total Cash Flow from Operating Activities</td><td>2,000</td></tr>
<tr><td></td><td></td></tr>
<tr><td>Cash Flow from Investing Activities</td><td></td></tr>
<tr><td>Equipment Investment</td><td>(15,000)</td></tr>
<tr><td></td><td></td></tr>
<tr><td>Cash Flow from Financing Activities</td><td></td></tr>
<tr><td>New Capital Lease</td><td>20,000</td></tr>
<tr><td>Term Loan</td><td>(15,000)</td></tr>
<tr><td>Line of Credit</td><td>21,500</td></tr>
<tr><td></td><td></td></tr>
<tr><td>Total Cash Flow from Financing Activities</td><td>26,500</td></tr>
<tr><td></td><td></td></tr>
<tr><td>Total Cash Flow
Opening cash balance (1/1/xx)</td><td>13,500
1,500</td></tr>
<tr><td></td><td></td></tr>
<tr><td>Closing Balance as of 3/31/xx</td><td>$15,000</td></tr>
</table>

We have just completed the Cash Flow Statement. This is the third key financial statement. It detailed how the cash was generated and how it was spent. It is the result or reflection of the operating policies of the company.

Planning Cash Flow

In Chapter Ten, Financial Planning, two planning methods were presented, a quick estimate method ("Change in Sales") and a detailed method. The detailed method will be the concentration here. (You may want to have those statements from Chapter Ten available to refer to.) Under the detail method, we determined the sales through earnings annually and by month, after taxes. It is time to project the cash flow for the planning period.

Three projection statements are prepared. My methodology (after completing the earnings plan) is to plan the cash flow and generate the Balance Sheet afterward. I employ this sequence as the Balance Sheet is dependent upon the financial policies of the company, which are built into the cash policies of the company for receivables and payables, etc.

Cash inflow from operating activities

Collections

How do we plan for collections? Easy! Using your DSO or Days Sales Outstanding in collections formula, we will determine the average number of days to collect an invoice. For example, if the DSO is fifty-five days, then for planning purposes, let's use two months and take another 1% off for bad debts. So, what is sold in January should be collected (99%) in March. Or you may have determined that you will collect 50% of January billings, 40% of February billings and 10% of current month billings (the total less the 1% for bad debts). There are your collections. This experience continues for each month.

To complete the accounting cycle and to create the A/R balance on the Balance Sheet, the calculation would be prior balance, plus sales, less collections. That calculation determines the new balance on the balance sheet, plan or actual. It all ties together and is all accounted for; including the bad debts.

Disbursements

Disbursements take a bit more thought, just a bit. Materials are probably the most prominent disbursement item and the easiest. Materials can be detailed from when it is ordered—to when it arrives—to when it is to be paid, by major type or major vendor, or in toto. Thus, if a company receives the goods from vendor A with remittances of thirty days or ninety days, then the disbursement will be planned accordingly. Labor is usually disbursed within the month it is earned unless it is material enough to be isolated and is special. Expenses are usually separated into current month disbursements (as with business taxes) or the following month's disbursements (as with telephone charges) for payment. Thus, all costs and expense disbursements are planned.

All other expenditures are handled on an individual basis. For example, a term loan repayment is disbursed either on a monthly basis or per a schedule. Disbursements for capitalized leases, repayments to owners, tax deposits quarterly, etc. are also planned. These impact the Balance Sheet, disburse the loan payment, and reduce the loan balance liability. Again, my preference is to leave a revolving loan out depending upon the circumstance.

In planning, when all the flows are totaled by month, a shortfall can be made up from a line of credit, other borrowing or from other temporary or permanent cash infusions. If there is an excess, pay down the line of credit. If there is an excess over a period of time, and the line of credit is paid off, and investing comes into play, dividends may be funded or any number of options becomes available.

When this is completed on a monthly basis, a picture of the peaks and valleys regarding cash becomes clear. Thus, while your company may generate cash, at some point during the planning period the company may need to borrow from a bank on a short-term basis (or longer as the case may be) and repay it in another part of the planning period. Starting with the opening cash balance, all the excess or shortfall should be added to the balance monthly. Thus, the peaks and valleys are shown as well as the ability to repay. Or, other types of funding may be the answer. But the point is made as it is planned and revealed. Other business decisions will be considered via the "what if."

Finally, all the disbursements data, as with the collections data, ties into and creates a Balance Sheet. We have planned the Income Statement, the cash flow, and a Balance Sheet now. We can calculate any of the ratios we want, including those a lending institution requires as part of the covenants of any loan. (See the Balance Sheet in Chapter Ten, "Budgeting and Financial Planning: Fixed and Variable Accounts.")

Detail Special Weekly and Monthly CASH FLOW Report

In order to manage the business' cash flow, one must be aware of additional information. Namely, what will the cash flow be in detail over the next month, six weeks, or two months? It is something that can be detailed and planned with reasonably good accuracy. It is something that must be performed and monitored closely. It is the center of your daily financial business activity and it is pure raw dollars. Nothing sophisticated here. Cash in, versus cash out.

There are two basic sections of the report, cash inflows and cash outflows. It is spread over a short-time period. The period I recommend is five or six weeks as it fully encompasses a month and is not restricted to the calendar of a month. The week is usually how people work and probably get paid. This can be adjusted to any basis, depending upon how a company operates including the following:

1. Does the company disburse on a weekly basis, biweekly, first and third Friday of the month, 10th and 25th of the month, etc.?

2. Does the company pay its employees on a weekly, biweekly, or 15th and end of month?

Note: The answer to these questions should determine how the detail cash-flow forecast report should be set up. (A super precise report, incorporating "float," or when actual checks will clear is not recommended. The more conservative checkbook balance is recommended.)

A spreadsheet is created. Below is an example of such a spreadsheet. My style is to have the disbursements detailed below and the sum placed up on the top section, working as a summary. This spreadsheet can take any format, as long as it is understood. The goal is to place in each box or cell amounts that will be, or is estimated to be disbursed in a particular week. These amounts may be what is actually in house that will have to be paid, or in the form of invoices or based upon experience and business activity. Lines should be set up for the larger individual disbursements (payroll, large vendors, or loans) and for groups of expenses. It's a matter of style and comfort level of the company.

When weekly collections are input, it is complete. You now have a good idea of your opening cash balance. (What will be collected by the week? What will be disbursed? And, what is the cash balance for the next week?) You can then plan for any shortfalls or excess cash positions.

For shortfalls, we can delay disbursements by (1) planning for certain items to be paid the following week, (2) plan to have certain accounts asked to advance their payment earlier than their normal schedule, or (3) utilize a line of credit. If the company chooses to delay a payment, then one must consider advising the vendor or not. If the company chooses to advance some collections, politics must be considered, who is to be called (and definitely not the same ones if previously done, if it can be avoided). The company may choose to draw up the line of credit, if there is one, and it must consider how close it is to the limit, now and in the future. Does the company have a large enough line of credit? To determine this, compare the line to the expected requirements forecasted in the monthly Financial Plan.

If there is excess cash, are all the discounts being taken and are the discount annual rates of interest (review Chapter Seven) to be taken better than the company can get via investing short-term. Can we negotiate with other companies to offer discounts for early and prompt payment? How long will this excess cash position remain? (Again, in the monthly plan, incorporating cash flow provides the basis for estimating the future months and the peak shortfall and loan requirement).

XYZ Company, Inc.
Cash Flow Projection

Week End	16-Jun		23-Jun		30-Jun		7-Jul		14-Jul		21-Jul	
	Estimate	Actual	Estimate	Actual	Estimate	Actual	Estimate	Actual	Estimate	Actual	Estimate	Actual
Opening Balance	20,015	20,015	8,295	1,380	8,745		9,195	0	7,695	0	6,845	0
Collections												
Cash	0											
A/R	25,000	28,000	50,000	43,000	50,000		50,000		25,500		50,000	
Other												
Total Receipts	25,000	28,000	50,000	43,000	50,000		50,000	0	23,000	0	50,000	0
Line of Credit Borrowing	20,000	7,000		0	10,000							
Total Available	65,015	55,015	58,295	44,380	68,745		59,195	0	56,195	0	56,845	0
Disbursements (tot from bot)	56,720	53,635	49,550	38,673	59,550		51,500	0	49,350	0	54,750	0
Closing balance	8,295	1,380	8,745	5,708	9,195		7,695	0	6,845	0	2,095	0
Disbursements Detail												
Payroll	32,000	31,000	30,000	28,500	30,000		30,000		30,000		30,000	
Payroll Taxes	2,720	2,635	2,550	2,423	2,550		2,550	0	2,550	0	2,550	0
Workman's Comp												
Commissions	8,000	7,500							8,000			
Ins-Health							1,600					
Insur-Other							1,500					
Rent							3,000					
Vendor A	2,000	2,000							2,000			
Vendor B	1,800	1,800							1,800			
Vendor C	5,000	5,000			25,000							
Office Supplies	200	200									200	
Telephone	3,000	3,000							3,000			

Let's examine the chart on the previous page. First, look at the disbursements. There are primarily four groups:

- Payroll and taxes
- Direct materials
- Expenses
- Loans and other finance disbursements
- A purely alphabetical system can be employed as well

There is no direct reason for any grouping, just the logic of grouping the disbursements. This is not an accounting statement or an actual disbursement from the accounts payable ledger; it's simply a planning and tracking tool. Thus, good approximations are the goal. So on a weekly basis, ask yourself, what is the payroll that actually gets disbursed? When do you disburse the items withheld (plus the company portion of taxes, etc.)? Do the same for materials. When do you believe you will actually pay the bill? If all the known bills have been put into the system, an estimate of what will be paid in the future should also be made. This is based upon recent, current, and near future business activity. Guestimated orders, receipts, and disbursements can be made, as with the payroll. Expenses will be done similarly. Loans should be known along with the due dates. One should also add a miscellaneous line as there is always something unexpected.

I suggest not listing every account (if the volume is overwhelming), only the principal ones and those easily known. Others can be grouped based upon experience. So we have estimated the disbursements over the period.

Now for the collections! This is just a bit more difficult. What have your collections been by week in each month? List your deposits by week. That starts the picture. Eliminate all "non-normal business" related deposits (line of credit infusion of cash, personal infusions, cash from sales of assets, etc.). Thus, what is listed is the weekly deposits of collection from sales. List the sales by week or month. And, do your customers pay against an invoice or against a statement? Generate two patterns: (1) monthly collection against monthly sales, and (2) weekly collections against weekly sales (delayed by so many weeks). This can be graphed, or better yet, initiate a percentage relationship on a test bases: collections of the month following the sales, or two months following the sales, or whatever seems to be emerging. The average DSO over a one-three-year period is a good starting point. When the highest relationship is found, use it. (This is also used in the collection pattern for the monthly cash flow portion of the financial plan via the DSO calculation) Do it also for a weekly pattern, or use the month estimate and split it weekly when a pattern can be found.

For example, one company found that their average DSO was twenty-one days another company determined that their DSO was fifty-seven days and that within each month the collections pattern on a weekly basis were:

Week 1	1/6th of the month
Week 2	1/6th of the month
Week 3	1/3rd of the month
Week 4	1/3rd of the month.

Summary

We have examined cash flow on three bases: (1) Actual, (2) monthly and annual plan, and (3) one to two month detail outlooks. The actual was detailed on the Cash Flow Statement produced with the monthly Income Statement and Balance Sheet. The planned forecast was developed both on the quick Change-in-Sales estimating method (usually on an annual basis which is not really sufficient) and in the detail forecasting method (which is recommended to generate a high degree of confidence). This was also part of the forecast Balance Sheet plan development. Also developed was a quick detailed look of what a business could expect to disburse within a short forecast period (five or six weeks out) in sufficient detail to plan for actual weekly collections and disbursements. This five or six week cash flow projection also relates (1) to the plan's monthly amount to ensure that it will not exceed any line of credit limit and (2) to detail into the plan regarding future months. It also details those demands on the Line of Credit or the ability to invest excess cash for longer periods (one month, three months, six month, etc.) if the forecast warrants such decisions.

Thus, we now know how we spent our cash in the past, what we can expect over the planning period, and what should actually occur within the next six weeks.

SECTION THREE

Analysis

CHAPTER THIRTEEN —
FINANCIAL TOOLS

Seasonality
Break-even Analysis
Sustainable Growth Rate
Interim Reports

Seasonality

Most businesses have some variation between the months during a year's operation. Compare this to a business in which each month has the same business activity throughout the year.

For example, a company manufacturing roofing shingles and is located in the Midwest or Northeast will have an extremely large swing in business between the cold months and the warm ones. Not too many people will have roofing installed during the winter. On the other hand, in retail, sales between Thanksgiving and year-end will enjoy significant sales, sometimes in excess of 25% of the total year's business. Therefore, why not plan for it accordingly and review the impact on the business that operates as such (e.g., sales, labor, inventory, accounts payable, accounts receivable, cash, etc.).

The chart below details the calculation of seasonal numbers based upon two and a half years of actual data. The last six months is projected data that is based upon managements experience with the business.

Each month is charted, all "like" months are summed, and the three-year monthly totals are each divided by the three-year total. For example, May totals to 1,180, divided by 15,933, generates the percent of the year occurring in May, 7.5%.

Thus, the most recent three-year experience becomes the percent of the year's sales factor for each month. If something is definitely known that would impact a particular month or months differently than the latest experience, then adjustments are made accordingly. It would affect the particular month(s) total and the three year grand total; thus, having some impact on all the monthly percents. The data presented are linked to the detail financial plan appearing in Chapter Ten, Budgeting and Financial Planning.

XYZ Company
Net Sates
($ Thousands)

	Jan	Feb	Mar	Apr	May	Jun	Jul	Aug	Sep	Oct	Nov	Dec	TOTAL
Year - 2	176	285	355	404	303	263	448	381	567	478	392	389	4,441
	4.0%	6.4%	8.0%	9.1%	6.8%	5.9%	10.1%	8.6%	12.8%	10.8%	8.8%	8.8%	100.0%
Year -1	411	277	400	416	378	398	451	402	478	423	389	389	4,812
	8.5%	5.8%	8.3%	8.6%	7.9%	8.3%	9.4%	8.4%	9.9%	8.8%	8.1%	8.1%	100.0%
								F C					
Year - 0	463	507	570	556	499	541	644	550	600	600	600	550	6,680
			Actual Total through July				3780						
	6.9%	7.6%	8.5%	8.3%	7.5%	8.1%	9.6%	8.2%	9.0%	9.0%	9.0%	8.2%	100.0%
Totals	1,050	1,069	1,325	1,376	1,180	1,202	1,543	1,333	1,645	1,501	1,381	1,328	15,933
	6.9%	7.6%	8.5%	8.3%	7.5%	8.1%	9.6%	8.2%	9.0%	9.0%	9.0%	8.2%	100.0%

Break-even Analysis (B/E)

Owners and managers always like to know how much business (sales) must be done to break-even (usually at the pre-tax level). Breaking even means at what level of sales will my pre-tax earnings equal zero; no profit and no loss. It means that all costs and expenses equal the sales value.

A business puts together a financial plan for a period of time or it reviews the latest year's operation. Based upon this data, the company can determine what sales must be to break-even. My preference is that it should be based upon the Financial Plan, which is the future and it incorporates business assumptions and underlies potential investment expenditures. To provide an example, let's use the Financial Plan created in Chapter Ten. (See below) We have restructured the Financial Plan Income Statement into categories of Fixed, Variable, and Partially-Variable costs and expenses. I have so indicated by placing the F, V or P next to each account line. Some are subject to interpretation, but it depends upon exactly what is placed in each account.

In summary, total fixed expenses (including half of the P expenses) are $715,091 for the year. To achieve the $7,300,000 in sales (which, too, are truly variable), $6,118,996 must be spent. Thus, for each dollar of sales revenue generated, 83.8 cents must be spent, with 16.2 cents (%) generated as marginal profit, which contributes to the coverage of fixed expenses. So, in order to break even, how many sales dollar units must we sell at 16.2% variable profit contribution to cover the $715,091 of fixed expense? Answer: $715,091 / .162 = $4,420,109. The proof is offered in the schedule.

Break-even is $4,420,000 annually or $368,000 per average month or since $4,420,000 is 60.5% of the $7,300,000 sales estimate. If sales are level throughout the year, on July 10th all the fixed expenses should be covered with the balance of the sales generating the $465,913 in pre-tax earnings. Now we know what the break-even point is. However, there are other uses for this analysis.

If XYZ were to hire a clerk at $20,000, how much additional sales must be generated to cover the salary and still maintain the $465,913 forecast? 20,000 / .162 = $123,457 in additional sales above the $7,300,000 forecast.

If XYZ wants to earn $600,000 in pre-tax income what sales level must be achieved? First, consider the $600,000 as a fixed cost, and add it to the $715,091; the total is $1,315,000. $1,315,000 / .162 = $8,117,300 (rounding).

If XYZ achieves an efficiency improvement (labor and materials) of .5%, what will be the impact? (The variable gross margin increases by .5% to 16.7%) $715,091 / .167 or the break-even point declines to $4,281,985.

XYZ Company
2xxy Fiancial Plan
Break-even Analysis

	F,V,P	Tot 2xxy	%Sales	Fixed Items	Vari Items	BIE Point
Variability						
Net Sales	V	7,300,000	100.0%		**7,300,000**	
Cost of Sales						
Direct Materials	V	4,972,698	68.1%		4,972,698	
Direct Labor	V	865,252	11.9%		865,252	
Plant Expenses						
Utilities	F	55,279	0.8%	55,279		
Depreciation	F	25,944	0.4%	25,944		
Rent	F	41,424	0.6%	41,424		
Health Insurance	F	107,312	1.5%	107,312		
Shop Supplies	V	22,381	0.3%		22,381	
Payroll Taxes	V	127,192	1.7%		127,192	
Total Plant Exp		379,532	5.2%			
Total Cost of Sales		6,217,482	85.2%			
Gross Profit		1,082,518	14.8%			
General & Administrative Expenses						
Salaries (Incl. Pit. Mgr)	F	248,420	3.4%	248,420		
Payroll Taxes	F	26,951	0.4%	26,951		
Commissions	V	86,963	1.2%		86,963	
Advertising & Promo	F	26,000	0.4%	26,000		
Auto Expenses	F	18,000	0.2%	18,000		
Misc Txs & Licenses	F	3,500	0.0%	3,500		
Office Exp. / Postage	F	24,000	0.3%	24,000		
General Insurance	F	32,250	0.4%	32,250		
Repairs & Maint.	F	1,200	0.0%	1,200		
Dues & Subscriptions	F	2,400	0.0%	2,400		
Travel, Meals & Ent.	F	36,000	0.5%	36,000		
Telephone & Utilities	P	16,021	0.2%	8,011	8,011	
Miscellaneous	F	2,400	0.0%	2,400		
Total Gen & Admin		524,105	7.2%			
Income from Operations		558,413	7.6%			
Other Income & Deducts						
Interest Expense	F	(56,000)	(0)	(56,000)		
Bad Debts	v	36,500	0		36,500	
Total Other Income / Deducts		(92,500)	(0)			
Pre - tax Income		465,913	6.4%			

Totals: Fixed, Variable Costs and Expenses		715,091	6,118,997	
Variable Gross Profit			1,181,004	
Variable gross margin cost			83.8%	
Variable Gross Profit Margin			16.2%	
BREAK-EVEN SALES		715,091	16.2%	4,420,106
PROOF				
LESS: Variable cost of sales @ 83.8%				3,705,015
LESS: Fixed Costs and Expenses				715,091
Total Variable and Fixed Costs and Expenses				4,420,106
Pre-Tax Earnings				0

Sustainable Growth Rate (g)

This is a percent growth rate that will represent the maximum (sales) growth a company can achieve from internally generated cash sources. Internal sources eliminate any form of borrowing (long-term, short-term, leasing, etc.) or any infusion of equity funds from consideration. If additional growth beyond the sustainable growth rate is desired, then the amount of funds must be identified as well as the source.

The Formula:

$$g = \frac{\text{net income}}{\text{Sales}} \times \frac{\text{sales}}{\text{assets}}$$

$$(1-(\text{total debt}/\text{total assets}))$$

This says: the sustainable rate of growth is equivalent to the net earnings percent times the asset turnover all divided by the inverse of the debt ratio. In plain English, per below, it is the return on owners common equity (this excludes any preferred equity) multiplied by the dividend retention rate* OR

$$\text{"g"} = \frac{\text{net income}}{\text{common equity}} \times \text{ the dividend retention rate}$$

If dividends disbursed are 40% of net income, then the retention rate is 60%. This reduces the return on owner's equity to that which is retained in the business.

Interim Reports

If the Accounting Department is operating properly, then you are getting financial statements on the 5th work day of the following month, or the 10th workday. Then SURPRISE, you didn't do as well as you thought; or you did better, or whatever. This surprise factor is relatively easy to fix.

Via a daily and month-to-date, or a weekly and month-to-date report, the manager can have certain important statistics that give him or her sufficient current data to manage the business or department. This usually consists of a dozen or so statistics, both financial and operational. For operations it usually consists of:

- Sales (the total of actual invoices process that day or week)
- Securements (new orders received for future sales; tomorrow or next month)
- Backlog (current value of orders plus new securements less sales)

- Units produced (product X and Y)
- Labor hours

For financial data, it might include:

- Collections
- Disbursements
- Checking account balance
- Accounts Receivable Balance (should match current balance plus sales less collections)
- Accounts Payable Balance (current balance plus new invoices less trade disbursements)
- Inventory (current level plus new production, less shipments)
- Line of credit balance

The list can go on and on and become more complex, but that is not the purpose of the report. It should be easily produced and have sufficient statistics to keep management informed without being overwhelmed on a daily basis. This report together with the weekly six-week cash flow should keep key members of the company up-to-date on how the company is doing. Depending upon the data incorporated, not all the recipients may receive the full report; and depending upon time, not all the columns may be available for reporting.

A sample of such a report is attached. In this example, bottom totals are for those that can be reflected accordingly, others are balances as of a date and do not lend themselves to totals. Some data (A/R invoices are not shown, production units are valued in the report) is hidden.

Remember it is an interim management report, not an accounting report. While accuracy is always required, the same accuracy level may not be attainable within this report as regarding the time necessary for its production, compared to the month-end financial reports.

XYZ company
Anytown, State
Daily Management Report
Month:March 2xxx

Date	Day	Net Sales	Plan	New Orders	Backlog	Units Prod	Collections	Disburse	Ck Bk Bal	A/R	A/P	Inventory	Line Credits
1-Mar	Sat				$353,000				$300,000	$1,500,000	$750,000	$1,250,000	$0
2-Mar	Sun												
3-Mar	Mon	$42,000	$60,000	$87,000	398,000	123,000	$23,000	$15,000	308,000	1,519,000	820,000	1,290,500	0
4-Mar	Tue	58,000	120,000	12,000	352,000	125,000	32,000	0	340,000	1,545,000	862,000	1,324,000	0
5-Mar	Wed	63,000	180,000	45,000	334,000	119,000	87,000	18,000	409,000	1,521,000	907,000	1,352,000	0
6-Mar	Thu	67,000	240,000	64,000	331,000	117,000	98,000	125,000	382,000	1,490,000	845,000	1,377,000	0
7-Mar	Fri		300,000										
8-Mar	Sat												
9-Mar	Sun												
10-Mar	Mon		360,000										
11-Mar	Tue		420,000										
12-Mar	Wed		480,000										
13-Mar	Thu		540,000										
14-Mar	Fri		600,000										
15-Mar	Sat												
16-Mar	Sun												
17-Mar	Mon												
18-Mar	Tue												
19-Mar	Wed												
20 Mar	Thu												
21-Mar	Fri												
22-Mar	Sat												
23-Mar	Sun												
24-Mar	Mon												
25-Mar	Tue												
26-Mar	Wed												
27-Mar	Thu												
2 Mar	Fri												
29-Mar	Sat												
30-Mar	Sun												
31-Mar	Mon												
Totals		$230,000		$208,000		484,000	$240,000	$158,000					

Notes;   Monthly gross sales plan $1,260,000; Production plan; 2,583,000 units

CHAPTER FOURTEEN — PRODUCT COST

Product cost is the basis for the first level of earnings and answers the basic question: "Am I making money on the product?" It is the intention of this chapter to provide an understanding of how the product costs are created. It is not expected for the reader to become a cost accountant.

Product cost generally includes the cost of materials, the cost of labor to produce the product, and other "plant" expenses directly associated with the production of the product (into inventory and on the shipping dock). REMEMBER: SALES LESS PRODUCT COST EQUALS GROSS PROFIT. Gross profit is before any overhead (i.e., selling, general and administrative expense, etc.). There are several methods to interperet and apply the overhead. There are approaches that may be unique, but it may be the way the owner (or the industry) wants the data to be reflected. Let's start with what one would consider a reasonable approach for an initial, quick discussion for a .com company:

> I once performed financial consulting for a company whose "product" was website development and the operation of a "community" for its customers; a completely different cost concept from a brick and mortar manufacturing or distribution business. Its general cost structure was actual website development, approximately 7% of net sales, with selling expenses of approximately 59%, and general and administrative expenses of approximately 26%. Pre-tax earnings were 8%. Further, this company had two major lines: new low-cost introductory products and more intricate products resold or upsold to the existing customer base, 59% and 41% respectively.

The website development costs for the lower cost introductory products were nearly nil. However, the associated selling expense, incorporating W-2 sales personnel, commissions, and advertising totaled nearly 77% of the sales of that particular group of products. The G & A charge was a proportional 26%. Thus, the "introductory" products were losing approximately 3% pre-tax.

The more intricate "resale" products incurred nearly all the web development expense, 41% selling expense (principally independent contractor commissions) against its sales and a proportional 26% charge for the G & A expenses. The result for this group of products generating is an approximate 26% pre-tax profit.

As you have surmised, the key here was to properly assign the various line items of cost or expense to the various products (splitting in detail certain line items of expense as necessary). It was an eye-opener for management. It evolved from the detailed Financial Plan and the restructuring its Chart of Accounts. Basically, leading to financial planning and control, and then to product costs and profitability.

In a traditional manufacturing and distribution brick and mortar environment, the cost basics are direct material, direct labor, and plant expense.

1. <u>Direct material</u> is material that becomes a part of the finished product and its packaging and ends up in the finished goods inventory.

2. <u>Direct labor</u> are wages and salaries used to produce and package the product that goes into the finished goods inventory.

3. <u>Plant expenses</u> are expenditures that are incurred to operate the plant, but may not be directly associated with the actual production of the product into inventory.

Note: The employees who actually manufacture the product are direct labor, as above described. However, those employees who support the product's manufacture such as warehouse personnel are considered as indirect plant expense. Subsequently, if the inventory is then assembled and crated or wrapped for shipment, this would not be considered a direct expense but an indirect shipping expense. Both the labor and materials, which are used for the crating or wrapping, would be considered as such. Repeating, this would probably end up as part of the plant expense. Let me state that these are general rules and that deviations occur all the time and are generated by company culture.

When a facility is solely devoted to product production, all of the common plant (indirect) expenses may be included in the product's cost, such as:

- hourly employee wages not directly associated with production (receiving, shipping and maintenance labor)
- shipping materials
- maintenance materials
- utilities
- production space rent
- the plant manager's salary
- facility costs including equipment and supplies
- various insurances

However, this is subjective in a facility containing the total business and other departments, such as sales and marketing. If that is the case, then an allocation of cost usually occurs. There is nothing wrong with an allocation. An allocation could represent the portions that will "not be incurred" if the department were not there. Again, ask the reverse question. What would happen to the cost of the product if this (a department, a machine, or an operation) were not here? The calculation would generate a reasonable answer.

There are two other items we should address in this plant expense scenario: (1) plant payroll expense and (2) contract items. Payroll expense consists of the company paid portion of Social Security and Medicare, Federal Unemployment (FUTA), and State Unemployment. These are generally placed into plant expense, including the payroll expense for "plant indirect" labor and direct labor. The direct labor payroll expenses are often included in the plant expense rather than included in the direct labor category. Thus, one should be aware of the combination of payroll expenses for all these activities and separate or combine when necessary.

Workers' Compensation may also be handled in plant expense as well as certain other insurances; it depends upon the company's accounting policy.

As an exercise, one may want to detail all the elements of direct production labor. All that is necessary is to create a simple support schedule that details the direct labor and then adds the expenses for payroll taxes (detail), and other expenses not usually obvious, such as the company's portion of health insurance, and other benefits, such as the company's portion of a 401k Plan. Remember, should a schedule be developed (see example below), items such as health insurance is a monthly fee, not hourly. Thus, I advise doing all the numbers on an annual basis, taking all the payroll expenses as a percent of the direct costs to develop how much a true direct hour of work cost is.

This is a supplemental schedule for management. It states that for every dollar spent on direct labor, other expenses are directly associated and must be incurred and accounted for on the company books. The result is the true cost of an hour of direct labor work.

XYZ Company Cost of Direct Labor					
	Rate/hour	Max Basis Cap	Annual	Percent	Rate/ hour
Straight time rate, 2000 hours	$14.00		$28,000.00	100.0%	14.00
SS	7.65%	$73,000	2,142.00	7.7%	1.07
Medicare	1.45%	no limit	406.00	1.5%	0.20
SUI	0.80%	$7,000	56.00	0.2%	0.03
FUTA	0.30%	$15,000	45.00	0.2%	0.02
Workers Compensation	6.00%	annual	1,680.00	6.0%	0.84
Company paid health insurance	$200.00	per month	2,400.00	8.6%	1.20
Total Benefits Cost per direct hourly employee			$6,729.00	24.0%	3.36
Total cost per hour			$34,729.00		17.36

No overtime is planned; although if it is, then at the 150% rate, add the hours and cost. Should your company have a stock purchase or 401k plan, then it, too, should be added. If your company provides the cost of coffee, paid holidays, etc., these costs should also be included.

Contract Items

Contract items are those operations sent to another business to perform. This is done for various reasons. Usually no in-house expertise is available or there is not enough volume to make it worthwhile for an investment. If it is done periodically, it can be handled in the plant expense category. However, if it is a regular portion of the product (such as anodizing certain parts in a product), then one must consider it as a direct material or as a separate direct cost item (such as "contract work"). The cost for the contract should be relatively significant.

Let's now cost out a product or products.

Direct Materials

Direct materials can be one material or several. For example, if you're making pants, the shapes are laid out on layers of fabric and cut. There is waste. Much the same as when a product is cutout from sheet metal. Thus, so much material goes into the production and so much material comes out as a good finished product. The result is that each piece is costed for not only its actual area of material, but also proportionally for the waste. Or, if we're introducing raw materials by weight, how much comes out versus how much goes in, individually or in combinations with other raw materials. ALSO, can the waste be recycled without deteriorating the material? Example, a product takes three sets of nuts and bolts. However, the experience is that for each 100 units, 315 sets of nuts and bolts are used. Why? Can it be due to lost items, or theft, or breakage? How do we then, determine the "true" material cost? We can determine by actual experience and by engineered usages (those calculated by engineers for the first time or as a result of proposed efficiency improvements). Then, this usage is extended by the cost of the raw material(s) to come to a dollar value of the material in the product.

Direct Labor

This is similarly done! How many hours of labor are required to produce the product? It can be calculated for one unit or for 100 units or whatever works best for your company. Essentially, it is the number of direct labor people needed to operate a machine versus the throughput (machine speed sensitive). Again, this is based upon experience and on engineered estimates and then adjusted for downtime, startup, and shut down (non-productive time, etc.). Carrying this a little further, if it is determined that twelve men at $10.00 per hour will produce the required 12,000 units in a year, then, the total cost is 12 x $10.00 x 40 hours per week x 50 weeks (if that will be the operation). The total divided by 12,000 units equaling $20.00 direct labor per unit produced. Also to be considered is employee payroll expense, which is generally placed within the plant expense. If there are two processes or more, and each has different production rates or machine speeds with different staffing, then each must be calculated for every item contained within the product. Thus, the cost of labor, labor rates, total staffing, and machine speeds are extended to total the dollar cost of a unit of production.

Plant Expense

This is where the fun begins. The schedule below is an example of a plant expense statement. A Plant Expense Statement is a set of annual expenses based upon experience or a budget, or it may be the expense being incurred throughout the period (several months or a year). There may be different budgets for several areas of an operation. However, for this exercise, there is only one budget for the plant. The simple way is to divide the total annual budget by the number of operating hours to determine the cost per operating hour. Then divide the cost per operating hour by the machine speed (units per hour) to get the rate per unit. This is a dollar item, so no further conversion is necessary. Thus, complete product cost includes:

- Direct material
- Direct labor
- Contracts if applicable
- A portion of plant expense charged to the product.

Voila, done!

XYZ Company Standard Cost Development Widgets: Green		
Plant Expense for the year ended December 31, 2xxx (for 50 weeks, 2,000 hours)		
Total Units		2000
	Fix/Var	Total Year
Depreciation	f	$25,000
Equipment Rental	f	2,000
Health Insurance	f	72,000
Indirect Wages	v	168,000
Maintenance	v	12,000
Payroll Expense	v	45,400
Rent	f	360,000
Salary: Pit Super.	f	60,000
Shop Supplies	v	10,000
Total Expense		$754,400

However, if there is more than one operation for more than one product assigned, on a scientific basis, the right amount of plant cost to each product or to a machine to develop as true a cost as possible. Thus, we would have a correct rate for each element of a product and for a product as a whole. **This is Activity Based Costing**.

The basic plant expense schedule is expanded to enable each operation of the plant to incur the correct amount of cost to each product. To do this, it is necessary to determine the bases for the assignment or allocation of each line item. Some of the bases might be:

1. Square footage or each operation as a percent of the total plant square footage,

2. Headcount,

3. Labor hours, etc., directly assigned (as with depreciation) which is specific to each machine.

Thus, the cost estimate is reassigned to each operation by line item. It totals to the amount assigned for each operation "exactly." The total then is assigned to each product again on a machine speed or goods throughput or operating hour basis, etc. for the products manufactured on that machine.

XYZ Company
Plant Expense Statement
For the Year Enued

Basis for Distribution		Total	Machine 1	Machine 2	Receiving/ Whse
Machine	a	2	1	1	
No. employees	b	31	10	14	7
%			32.3%	45.2%	22.6%
Direct labor	c	$651,000	$280,000	$371,000	
%			43.0%	57.0%	
Operating hours	d	4,000	2,000	2,000	
%			50.0%	50.0%	
Machine speed (ft/min)	e		10	15	
%					
Square footage	f	38,000	8,000	20,000	10,000
%			21.1%	52.6%	26.3%
Specific	g				

Plant Expense Statement

Distribution	Basis	Total	Machine 1	Machine 2	Whse & Office
Depreciation	a	$25,000	$ 10,000	$ 15,000	
Equipment Rental	g	2,000			2,000
Health Insurance	b	72,000	23,226	32,516	16,258
Indirect Wages (Receiving/Whse Labor)	g	168,000			168,000
Maintenance	f	12,000	2,526	6,316	3,158
IndirPayroll Expense	Indir. Labor	45,400			45,400
Rent	f	360,000	75,789	189,474	94,737
Salary: Plant Superintendent	h	60,000			60,000
Shop Supplies	c	10,000	4,301	5,699	
Total Expense		754,400	115,843	249,005	$389,553
Sub total Mach 1 & 2		364,847	31.8%	68.2%	
Reallocation of Warehouse & Other		389,553	123,877	265,676	
Basis: Mach 1 & 2					
Revised Totals			$239,720	$ 514,681	
Cost per operating hour	2,000 hours		119.86	257.34	
Cost per unit			0.20	.286	
Units per hour			600	900	

We now have the correct cost for each part of the operation to construct a product and for the total product, going into inventory. And we can now determine whether a profit is made on each sale and whether it is sufficient. **Selling price minus product cost equals gross profit. Gross profit divided by the selling price (all per selling**

unit) will produce the profit margin. This is the initial part of whether the profit is sufficient. Remember, no selling, general and administrative expenses, nor other income and expenses, have been allocated yet.

	$	%
Selling price / unit	$125.00	100.0%
Product cost / unit ("scientifically" per the above)	82.50	66.0%
Gross Profit	42.50	34.0%

CHAPTER FOURTEEN A —
STANDARD COSTS (AND VARIATIONS)

Standard costs are product costs that are estimated to be achieved at the beginning of the forthcoming budget year. They are the above product costs that are adjusted for changes in all areas. The above product costs are generated from both usage of the various direct materials and direct labor and the price of each. Both are estimated to be in effect, generally, at the beginning of the forthcoming year. Material costs are estimated from past experience or from notices received from the vendor. Usages are adjusted for equipment investments, staffing changes or formula changes, and other efficiencies. The indirect expenses (plant expense) are adjusted for the new budget and any reassignment.

Standard costs are used (1) to value the cost of product into inventory and (2) to provide a good estimate of full product cost to determine product profitability, depending upon the selling price. Then comes analysis. How well did the product do versus the standard?

<u>Example</u>: The XYZ Company manufactures widgets for the sailing industry. They produce several products at their plant in New Bedford, Massachusetts in a simplified in-line operation. The production facility consists of one building, in good repair in the industrial section of town, where there is good access to all distribution. The facility also has all the shipping, receiving, and warehousing necessities to operate, as well as a small repair shop to address whatever "normal" repairs are necessary to maintain the operation. Sales, marketing, and general management are located in another facility in the commercial business section of town. The company wants to detail the actual cost of each widget and to install a standard cost system for current and future operations.

Harvey Goldstein

The following example details the development of the standard cost of 100 green widgets (the unit of measure). It's been determined that there are three groups of elements comprising the cost: (1) direct materials, (2) direct labor, and (3) plant expense. Using the same data as above, the plant expense costs are examined first.

The plant expense statement initiates from the annual budget to an "average" month, then to a month adjusted for actual production. This "Volume Adjusted Budget" column adjusts the variable expense items to what they should be based upon, the actual volume of production. Then the **actual** expenses are placed in the next column. At this point, a difference can be taken by line (between the actual and what it should have been on a adjusted budget basis) or in total (i.e., $65,613 vs. $61,583).

Note: I have set the units equal to 100 pieces, which matches the net line speed of the production equipment. Thus, units and time are the same for simplicity.

<table>
<tr><td colspan="6" align="center">XYZ Company
Standard Cost Development
Widaetsareen</td></tr>
<tr><td colspan="6">Plant Expense for the year ended December 31, 2xxxx (for 50 weeks, 2,000 hours)</td></tr>
<tr><td>Total Units</td><td></td><td>2,000</td><td>167</td><td>190</td><td>190</td></tr>
<tr><td></td><td>Fixed/
Variable</td><td>Total Year</td><td>Per Month</td><td>Volume Adjusted budget</td><td>Actual</td></tr>
<tr><td>Depreciation</td><td>f</td><td>$25,000</td><td>$2,083</td><td>$2,083</td><td>$2,083</td></tr>
<tr><td>Equipment Rental</td><td>f</td><td>2,000</td><td>167</td><td>166.67</td><td>0</td></tr>
<tr><td>Health Insurance</td><td>f</td><td>72,000</td><td>6,000</td><td>6,000</td><td>5,800</td></tr>
<tr><td>Indirect Wages</td><td>v</td><td>168,000</td><td>14,000</td><td>15,960</td><td>13,500</td></tr>
<tr><td>Maintenance</td><td>v</td><td>12,000</td><td>1,000</td><td>1,140</td><td>850</td></tr>
<tr><td>Payroll Expense</td><td>v</td><td>45,400</td><td>3,783</td><td>4,313</td><td>3,500</td></tr>
<tr><td>Rent</td><td>f</td><td>360,000</td><td>30,000</td><td>30,000</td><td>30,000</td></tr>
<tr><td>Salary: Plant Super.</td><td>f</td><td>60,000</td><td>5,000</td><td>5,000</td><td>5,000</td></tr>
<tr><td>Shop Supplies</td><td>v</td><td>10,000</td><td>833</td><td>950</td><td>850</td></tr>
<tr><td>Total Expense</td><td></td><td>$754,400</td><td>$62,867</td><td>$65,613</td><td>$61,583</td></tr>
<tr><td colspan="6">Note: Payroll expense includes the taxes tor all employees: warehouse, receiving, superintendent, and direct labor</td></tr>
<tr><td colspan="2">Hours of Operation</td><td colspan="4">2000 (50 weeks at 40 hours/week)</td></tr>
<tr><td colspan="2">Hours per average month:</td><td colspan="4">166.67</td></tr>
<tr><td colspan="2">Cost per unit (operating hour):</td><td colspan="4">$377.20 ($754,400/2000)</td></tr>
<tr><td colspan="6">Machine Speed: 100 units per hour (Average rate including startup, shut down and unscheduled downtime)</td></tr>
<tr><td colspan="2"></td><td colspan="4">Actual</td></tr>
<tr><td colspan="2"></td><td colspan="3">Operating Hours</td><td>200</td></tr>
<tr><td colspan="2"></td><td colspan="3">Production (100 Widgets)</td><td>190</td></tr>
<tr><td colspan="2"></td><td colspan="3">Rate per Hour</td><td>1</td></tr>
</table>

Here are the details for Direct Material and Direct Labor and the Overhead to complete the Costs Analysis:

XYZ Company						
	Standard Cost: 100 Green Widgets			Actual Experience		
Direct Material	Quantity Price Unit	Price	Cost	Quantity	Price	Cost
Iron Bars 10 lbs, 2" x5'	20.000 Bar	$15.00	$300.00	20.000	$14.00	$280.00
Paint green,specs	10.000 gallon	10.00	100.00	9.000	10.00	90.00
Boxes 2" x 10"x5'	20.000 box	1.00	20.00	20.000	1.00	20.00
Total DirectMaterial			420.00			$390.00
Direct Labor	10.000 work hours	$14.00	140.00	10.526	14.84	156.21
Plant Overhead	1.000 Oper.Hours	$377.20	377.20	1.053	307.92	324.12
Cost per 100 Widgets			$937.20			$870.33

Note: For this example, no separation is made between the bar cutting and machining operation and the painting operation. We are not that sophisticated (here) at this point for efficiency, cost control and analysis of each. But it can easily be accomplished by preparing budgets, staffing and efficiencies for each.

To verbalize the detail: the product consists of three raw materials (iron bars, paint, and packaging boxes), direct labor, and plant overhead. The direct materials cited above are further detailed to price and materials required, per the company's historical experience and the purchasing department's input regarding material price as of the beginning of the forthcoming period. To produce a unit (100 green widgets one foot in length), it requires 20 units of 5-foot long bars (2" in diameter) at an estimated cost of $15.00 per bar (as of 1/1 of the new year); plus 10 gallons of green paint at $10.00 per gallon and 20 boxes to package the 100 bars at $1.00 per box. When multiplied out, the estimated standard cost of the materials totals $420.00. Direct Labor costs $140.00 for the 100 bars and plant overhead costs $377.20. Total cost is $937.20 for 100 green widgets.

The **direct labor** is developed as follows: this is a continuous operation; ten men are required to actually operate the production equipment. The operation consists of inputting the five-foot bars into the equipment. The operations within the line include (a) cutting the bars to lengths of one foot each, (b) machining the bars, (c) painting the machined one-foot bars, (d) drying, (e) take-off and (f) boxing. The equipment averages 105 lineal feet per hour after adjustment for startup, shutdown, and estimated unplanned downtime (per eight hour shift). This line speed will produce 100 units per hour. Staffing consists of ten men on the production line, all men are cross trained and all get the same pay of $14.00 per hour.

How do we know that 100 widgets are produced? Production reports are generated every day by plant personnel, and then reviewed and summed by the plant superintendent. This includes the number of bars in, the number of widgets outs, the number of gallons of paint used, pounds of scrap metal from the machining, and any other statistics desired. This is the basis for confirming the machine statistics and the engineers' forecasts. It also confirms what must be reported to accounting. This is also part of the basic internal reporting and management of the company. Thus, over a period of time we know that for an eight hour shift, with all the starts and stops and repairs, the line takes so much input and generates so much output.

Plant Expense

Plant expense is detailed per the above annual estimated Expense Statement for the forthcoming year. This is based upon the experience plus any known adjustments for changes for the forthcoming year. This experience could be since the beginning of the current year and an estimate of the balance of the fiscal year, or the last 12 months or longer if it is still relevant. One specific item within the plant expense budget is the staffing of all indirect labor (seven employees for 2000 hours at $12.00 per hour equals $168,000). These employees will handle all the materials receiving, shipping, warehousing, and repairs. This is usually validated as a percent of sales vs. the prior period. In the short-term, staffing is usually static. However, in the "long-term," it would increase or decrease with demand and sales (hours or people). Review the chapter on Budgeting for more information.

The total of the plant expense budget is $754,000 for the year, $62,867 per average month, $377.20 per average operating hour (for the estimated 2000 operating hours in the year). At the production rate of 100 good widgets per hour, it will take one operating hour to produce them. Thus, the calculation is made of 100 per operating hour. If production were at 250 widgets per hour, then it would take 4/10 hours to produce one 100 widget costing unit.

For direct labor, the concept is much the same. The production operation requires ten employees to operate the equipment. At $14.00 per hour for each of the direct labor staff, the hourly direct labor cost is $140.00. Again, for one 100 unit cost, 10 man hours (one hour for the 10 operating employees) will be the basis. If productions were at 250 widgets per hour (2.5 units), then each unit would be produced in 0.4 hours with the 10 laborers, $56.00 per unit, if that were the basis.

The cost, or standard cost of 100 green widgets has now been determined or calculated.

Variations

The next step is to compare how we actually did to our historical experience and the standard cost we have developed. The data is detailed on the schedule above wherein the actual experience has been detailed. This data is based upon the accounting system and through production reports. Plant expenses for the month total $61,583 (based upon the accounting for the month's actual expenses); the plant operated for 200 hours in the month (20 days at 9 hours per day and 2 days at 10 hours) and produced 19,000 widgets (190,100 units of widgets). The calculation of actual cost includes <u>material usages</u>:

1. 19,000, 190 units, 3,800 bars were used. The warehouse records will confirm the data submitted by production regarding bars input into the system (including green paint and boxes). From a warehouse standpoint, opening inventory, plus receipts, less closing inventory equals the number of units that are left in the inventory (confirmed by signed warehouse receipts).

2. 1,710 gallons of green paint were used to produce the 19,000 widgets. (The validation is the same as in "A.")

3. 3,800 boxes were used; again validated.

Material Purchase Prices: Material purchase prices are those prices paid for the materials purchased (actually received) in the month.

Direct Labor Usage: Direct labor usage is the number of manhours divided by the number of units produced: 200 operating hours x 10 men per shift (all staff there for each operating, no absentees) / 190 = 10.526.

Direct Labor Price: Direct labor price is gross wages for the months for the 10 direct labor employees. In the scenario, there were 20 days at 1 hour overtime and 2 days with 2 hours overtime for each of the 10 direct labor employees. Overtime is at $21.00 ($14.00 x 1.5) per hour. Thus total wages would be $29,680; total man hours are 2,000. Thus the average cost per man-hour is $14.84.

XYZ Company Calculation of Direct Labor Cost						
	Days	**Hours**	**No. Staff**	**Rate/Hour**	**$**	**Hours**
Straight	22	8	10	14	24,640	1,760
OT	20	1	10	21	4,200	200
OT	2	2	10	21	840	40
Total					$29,680	2,000
Per hour					$14.84	

Plant Expense Usage: Cost (price) per hour equals the actual plant expense statement totals $61,583 for the month; divided by 200 operating hours is $307.915 ($307.92 rounded) cost per hour. The plant used 2,000 man-hours of labor (200 operating hours with a ten-person staff) to produce 190 widget units; the usage is 1.053. It could be noted that the principal reason for the below budget plant expense is that indirect labor was light due to employee absentees, if that were the case.

Variations from Standard Cost

There are six variations to be calculated. These are the definitions:

1. Material Price: The difference caused by a change in the actual material prices versus the Standard or estimated price. This is based upon incoming materials purchased (actually received into inventory). Thus, the variation is not directly associated with the material put into production.

2. Material Usage: The difference between the estimated material usage and the actual material used in production.

3. Direct Labor Price: The difference between actual and planned labor price wages. It could be the result of raises or overtime greater than anticipated within the "standard price" or a different combination of personnel with different

wage scales than those calculated into the rates paid for the positions to be staffed, etc.

4. Direct Labor Usage: The difference between the amounts of labor estimated to produce one unit of widgets and the actual amount. This is the result of staffing differences or "machine speed."

5. Expense Volume[1]: The difference between the amount to be absorbed into inventory via production (the standard plant overhead, dollar cost per unit, times the units produced) and the planned amount to be absorbed (volume adjusted Plant Expense estimate, see calculation).

6. Expense Expenditure[1]: The difference between the volume adjusted Plant Expense budget amount and the actual spending.

Using the descriptions above, here are the calculations:

---

[1]   These last expense item analyses boil down to the differences in spending between plan and actual. However, they are further split via the formula as to their causes: volume or just spending.

XYZ Company
Standard Cost Variation Calculation
Widgets: green

	Actual	Standard	Units	Purchased	Dollars (Fav)/unfav
Material Price					
Iron Bars	$14.00	$15.00	pieces	3600	(3,600.00)
Paint	$10.00	$10.00	gallons	1900	0.00
Boxes	$1.00	$1.00	boxes	4180	0.00
Total Material Price Variation					(3,600.00)

Note: Material price variation is calculated upon the purchased amounts within the month. In the calculation above, the amounts purchased did not match the amount relieved from inventory that went to production.

Material Usage	Actual Use	Standard Usage	Standard Price	Units Produced	Dollars (Fav)/unfav
Iron Bars	20.0	20.0	$15.00	190	0.00
Paint	9.0	10.0	$10.00	190	(10.00)
Boxes	20.0	20.0	$1.00	190	0.00
Total Material Usage Variation					(10.00)

	Actual	Standard	Units	Usage	Dollars (Fav)/unfav
Direct Labor: Price	$14.84	$14.00	Labor hours	190	$159.60

	Actual Use	Standard Usage	Standard Price	Units Produced	Dollars (Fav)/unfav
Direct Labor Usage	10.5	10.0	$14.00	190	$7.00

Plant Overhead	Volume Adjustment Plant Budget	Actual Production	Standard Cost	Absorbed into Inventory	Dollars (Fav)/unfav
Volume	65,613.00	190	$377.20	$71,668.00	(6,055.00)

	Actual Expenses		Volume Adjustment Plant Budget		Dollars (Fav)/unfav
Expenditure	61,583.00			65,613.00	(4,030.00)

There are several other levels of product cost, which are important to understand including marginal cost and full cost.

Marginal Cost

First, the definition. Essentially all fixed costs are eliminated. Thus, the only "costs" included are those which actually have to be incurred with the product, the **Variable Costs** which are the cash costs. Remember, the variable costs are direct materials, direct labor, and variable plant expense. A review of the plant expense reveals that the various expense items are designated "f," for fixed and "v" for variable. In the example, the total variable expenses are $235,400 budgeted for the year, $117.70 ($235,000 / 2000 planed units) standard for each 100 widget unit of production. Thus, the cash cost totals $677.70 (including direct materials and direct labor) at standard cost. At actual cost, if it were reflective of the year to date with no major adjustments foreseen, then the cost would be $546.21 ($98.42 plant, $390.00 materials, $156.21 labor). One other cash cost is the sales commission to the sales representative. (There may be others, one must use their understanding of the business to assess this.) If it's 10% of the sales (selling price per unit), then add that to this cost, simple!

Thus, if the company were to accept an order for $700.00 versus our regular price of $1000.00 per a 100 unit widget unit, then we would be paying $70.00 in commission. The total cash cost is $646.21, no matter when the disbursement is made. For example, commissions are likely to be paid in the month following the sale or later, but it will be paid and is a cash expense directly related to the sale. We are also assuming that our costs and production efficiencies remain consistent. If not, then the price must be adjusted accordingly. Plant actual variable expense is better calculated over several months rather than a one-month snapshot.

This is a level of cost that is used to a) secure additional business when all the fixed costs have been absorbed or "covered" by sales already made for the year-to-date, or b) to secure business at a low price for whatever reason and ensure that the direct cash cost of the shipment is covered with the price.

Full Cost

This concept is used to determine if a particular product is making a profit after all regular business expenses have been "charged." Thus, from the selling price, initially deduct the actual cost of the product (not just the standard, but all the variations). Next deduct other expenses, which are directly associated with the product. Most likely commissions, if there are different rates for different products. All other expenses left, which have not been assigned should be defined as a percent of sales. (What is left divided by the total of the company's sales). Apply that percent to the selling price of the product.

Define other expenses to include all General and Administrative expenses and Other Income and Deduction items so as to bring it to the pre-tax line. Thus, from the selling price (actually an average of all sales of that product), we have taken or charged the cost of the product, directly assignable expenses, and all other expenses on a percent of sales basis. What is left is the profit dollars of the product which should be expressed as a percent of the selling price. We have determined whether the product is making a profit, via the margin (percent of selling price) and whether it is acceptable. An example is offered below.

Per Unit Widget Profitability		
	$	%
Gross Selling price	$1,300.00	103.1%
less: R,D &A	(39.00)	-3.1%
Net Sales	1,261.00	100.0%
Cost of Goods Sold	937.20	74.3%
Gross Profit	323.80	25.7%
Selling Expenses	151.32	12.0%
General & Administration Cost	100.88	8.0%
Operating Earnings	71.60	5.7%

Since this is a one product line and a one sales force operation, the application of the Selling and General Administrative expenses (made up for this example) is simple. For a "multi-line" operation, reread the opening discussion of Product Cost describing a web site company for such an analysis, or assign the S,G & A expenses in a manner similar to spreading plant indirect costs.

CHAPTER FIFTEEN — COST OF CAPITAL; BUSINESS INVESTMENT ANALYSIS; CAPITAL BUDGETING

Your want to enhance you business' or department's profits. Let's be very pragmatic here, we are a capitalistic society. The goal may be simply to make more money, improve performance, or just to "maintain" the product or service. We can continue as we have always operated, or if there is an opportunity, we can do something to enhance the operation by changing location, enhancing the product, add new products, etc. Is it worthwhile? Will it "pay off"? What do we look at? How do we determine the possible pay off and how do we do it? How will the opportunity to enhance the operation or product fit into the whole picture? In this section, we address these questions.

In response to the above, we must examine several basics! First, what is the cost of money? Second, what will the project "return" be? Is it going to return to the investors sufficient earnings to warrant the investment? Third, how does the project fit into the total company investment picture; and the total business picture? REMEMBER, the overall goal of a business is to enhance the wealth of the stockholders.

Cost Of Capital

This section will discuss the "costs" and the "securities" or "instruments" that may be found in small and large companies. It will proceed to explain the cost and valuations of each, and sum them into the overall "cost of capital" or the cost of funding for a company (the average cost to raise investment money).

Cost of Capital for "small" Companies

Why do we want to know this? Because, if a project or investment returns X% (as we will find out below), it is important to know that the project or investment returns more than the cost of the funds invested in the project.

One means of raising money is for the owners or stockholders to invest their own money. So, if you were an owner or stockholder, is it worthwhile for you to do it?

An individual has his or her own goals. There are many investment alternatives including: savings deposited at a bank at x%; or investing into U.S. Treasury certificates with "absolute" safety, but at the lowest interest rates. There are risks with each alternative.

With a savings account, the bank may fail; hopefully, your account is fully insured via a Federal Agency, such as the Federal Deposit Insurance Corporation (FDIC). With a U.S. Government security certificate or bond, the Federal Reserve System (or commonly called the "Fed") may raise the interest rate. So while you still get the same dollar amount for each certificate you own, the certificate or bond's market value declines. This is because new bonds or securities have the same face value, but with higher interest rates. The face value of your bond hasn't changed, just the current market value, and it will be fully paid at its maturity. Also investing in private securities that are publicly traded is an alternative that carries its own analysis, risk, and return. There are many other opportunities in which to invest, each carrying its own risk and control.

Personal Investment: If one does choose to invest in one's own business, how much of a return is satisfactory? Also, investing requires funding, which has its own cost. So what is the cost of money to an owner? It could be the alternatives above (investing in U.S. Treasury securities or into a time deposit, etc.) or simply the return one is earning in their own business. What is the investor receiving for their money?

Lets start with the company's own "Return on Equity," Net Income divided by the Net Worth of the company. If nothing else, it reveals what is being earned on the money invested; defined as if one invests in a savings account. This return should be higher as the risk is more. However, what are the dividends being paid on the Net Equity invested? The dividends really determine the return. The return could be an actual dividend (or other benefits paid to the owner). In a sole proprietorship, or in a partnership, or sub-chapter S corporation, LLC, etc., it could be a dividend or simply a salary received (in theory, the dividend) against the Net Worth of the company. Thus, it is the total earnings divided by the investment (Net Worth).

Borrowing from a bank: What is the true cost? On an annual basis, it is the total cost of the loan divided by the amount of funds made available to the borrower.

A bank bases its interest rate upon the Federal Reserve Bank's basic interest rate. Thus, if the Fed's rate is 6.0%, then your bank may have a prime rate of 8.5%. The prime rate is what a bank charges its best customers. For other customers, a bank will probably charge "prime plus X percentage points." So, to an individual customer, the rate might be prime plus 1.5%, making the actual rate 10.0% per annum. Add to that a 1.0% origination fee; the effective rate to a customer is 11.0%. If a compensating balance is required, it increases the rate accordingly. A compensating balance is a sum of money (part of the loan) that is required to be in a customer's account at the bank at all times. So, if a 10.0% compensating balance is required on a $100,000 loan, the funds available to the customer are $90,000. If the above is true, then 11.0% of $100,000 is $11,000. This is the cost for the $90,000 available to the customer. The effective interest rate (or Annual Percentage Rate) is $11,000 divided by $90,000 or 12.2% if this is a one-year term loan (i.e., $100,000 borrowed and repaid 365 days later).

There are all types of loans with different terms (e.g., the example above, a three year term loan that calls for periodic repayments, etc.). Another is a Revolving Line of Credit loan that is renewable annually. A Line of Credit may have a "clean-up" provision, which means for a period of time the Line of Credit is required to be zero within a twelve-month period. Some banks require this to ensure that it is not a long-term loan in disguise.

In many smaller companies, a loan is probably going to be used or assigned for a particular purpose to buy specific equipment, expand inventories, etc.

Matching: One final point, the length of any loan should be closely matched with the project's length. It would not be wise to use a Line of Credit to finance a new machine that is expected to last five or more years. A three-year loan with periodic repayment provisions would be more desirable. Conversely, a three-year loan should not be used to finance short-term increases in working capital.

Cost of capital for publicly traded companies.

This outlines the potential cost of capital with a brief explanation of the different types of securities a company issues.

It should be noted that for larger companies, capital that is raised is generally not specifically assigned to a project, but rather put into the general operating funds for day-to-day operations to be disbursed as needed. It is repaid from that cash pool.

There are three major types of securities that a company issues: (1) Bonds, (2) Preferred Stock, and (3) Common Stock. In most cases, common stock is the overwhelming security that is publicly issued. Following are the major characteristics of each:

Bonds

Bonds are a debt obligation of a company, there are a variety of types. They are found in the Long-term Liabilities section of the Balance Sheet. Most companies issue bonds in denominations of $1,000 (Par or face value). They may be general obligation bonds or secured by specific assets including the plant, property, or equipment (such as, buildings, machines, railroad, or airline equipment). Bonds may be issued for ten or twenty years or however long the company wants. They carry an interest rate. Distribution of the interest rate usually takes place semi-annually. Some bonds are convertible into stock. (A "sweetening" feature that could lower the interest cost to the company).

Bonds have three other possible features:
1. Call provision: The issuer may reserve the right to call for the redemption of the bonds at a certain price. This usually happens when the market interest rate drops. This raises the price of the bond to arrive where the interest paid, relative to the increased price, matches the new market bond yield for this type of security. (See the bond example at the end of the chapter) It behooves the company to analyze the economics of refinancing the bond issue with lower interest bonds.
2. Sinking funds: These are funds set aside to ensure that at maturity, there will be cash available to redeem the bonds.
3. Serial Bonds: Serial Bonds are designed to be periodically repurchased (actually redeemed); thus reducing the outstanding securities as the issue matures.

Usually, part of the agreement (called the "indenture") that details all the obligations of the issuer within the bond requires that there be a trustee. A trustee is an individual or company appointed by the lenders to represent the bondholders in all regards. Generally, it is a bank or a securities firm or an individual who specializes in these matters.

Preferred Stock

Preferred Stock is an equity or ownership security. It is found in the Net Worth (Net Equity or Stockholders Equity) section of the Balance Sheet. In a liquidation of any kind, it gets paid (redeemed) after all the creditors get paid, up to its Par value. These securities get paid before the common stockholders. Some other key elements include:

1. The Preferred Stock Par value is meaningful (as just explained) and it generally is sold at Par or based upon it.
2. It generally has cumulative dividend rights, which means if dividends are missed, no common dividend can be distributed until all the back dividends owed to the preferred stockholders are fully paid.
3. Sometimes there is a convertible feature, which means that under certain conditions, these preferred shares can be exchanged to common shares.
4. If dividends are missed for a specified period, voting rights may be conferred upon the preferred shareholders.
5. It is usually sold as a dividend dollar and relative to the Par, it generates a percent yield, or it is stated as a dividend percent to Par. Dividends are usually distributed quarterly or semi-annually.

Preferred stock is a relatively secure stock, with a lesser risk than common stock as it has priority above common stock in liquidation and has a known return. However, the stock movement is not great as it is controlled by the dividend as a percent of Par. It will move if the company is in trouble or if the dividend as a percent of par is not in line with market expectations. (See discussion below on valuing preferred stock.)

Common Stock

Common Stock is the basis for ownership of any company, regardless of size, if it is a corporation. The Par value as stated on the certificates and on the books of the company is nearly meaningless. Thus, when common stock is sold by the company raising equity capital, from an accounting standpoint, the funds are accounted for on two lines or accounts, "common stock" at Par and "additional paid in capital" for those funds received in excess of the Par value or less than the Par value. It is the lowest claim on the company in any liquidation, and on dividends. The liability to the shareholders is limited to their purchase of the stock. The price is volatile with the potential for the greatest gains accruing to these shares, as well as the greatest risk. Also, dividends can grow and grow significantly or not at all; thus affecting the price. It is measured by earnings per share (EPS); that is net income less dividends to preferred shareholders, divided by the number of common stock shares outstanding, fully diluted.

Note: Diluted means the number of common shares used as a divisor, it also includes shares from other company securities that can be converted into common shares from preferred stock or bonds and stock options.

In a bankruptcy or other dissolution of the company, after legal and employee obligations are paid, the bondholders get paid. First, the secured bondholders are paid or take control of the specific assets pledged; and then, the general obligations bonds get paid, which are usually backed by the general faith and credit of the company. If there is anything left, the other general creditors are paid, then the Preferred Stockholders, and lastly, the Common Stockholders.

Valuation (of each type of security)

The following discussion is meant to outline the prominent costs of funding a company. It is not intended to provide an in-depth security analysis. The discussion centers on the cost of capitalization. The capitalization funds are those funds placed into the company for investment purposes, not as a result of operations as in accounts payable. As per above, the capitalization is in stock or bonds with each having its own risks and rewards. Bonds are a liability, rather than ownership. In general, the securities are traded as an investment in the marketplace.

Bond financing cost is determined by the interest rate on the bond. Thus, if a company has $2.0 million dollars outstanding in 7% bonds, then the cost is 7% for that portion of its capitalization. Simple! The cost of the funds to the company is $70.00 per year for each $1,000 bond. Let's not discuss here the maturity or the current price in the market. However, the company has the funds and is paying 7.0%, and it doesn't have to buy back the bonds; thus, the 7.0% is current, before taxes.

The interest on the bonds or debt securities is chargeable against earnings. It usually appears in the Other Income and Deductions section of the Income Statement; thereby, reducing pre-tax earnings. Thus, this cost is partially "offset" by reducing the income tax that the company must pay. Therefore, if the incremental tax rate is 40% for that level of company earnings, then the company benefits from the interest expense as it lowers the taxes that must be paid. So, if the interest rate is 7.0% and the interest expense lowers the taxes to be paid, then dollar for dollar, the actual cost is reduced by 40% (the income tax rate). Thus, the true cash cost to the company is 7.0%-(40% x 7.0%) or 4.2% (interest expense less reduced taxes resulting from the expense on the income statement).

Preferred Stock

Cost for Preferred Stock is determined by the formula: Dividend / Market Price. If the stock is trading at Par and is receiving the promised dividends, then this too is simple to determine. For example, a 9.0% Stock with a $100 Par value is, just that, a 9.0% cost. However, if overall interest rates are reduced by the Fed, the market may perceive the $9.00 (9.0%) as a very good return and may bid up the price from $100.00 to $110.00. In this case, the $9.00 annual dividend will not change, but a new investor would only be receiving 8.2% ($9.00 / $110.00), which is acceptable in the marketplace as that's what the market has made the price.

$$\text{Dividend / Interest rate} = \text{Price}$$

Common Stock

The valuation of this security is a bit different. There is a Par value. However, it does not have any particular meaning. The valuation is, generally, via the following formula:

% Return = (D/P) + g
Where: D is the current annual dividend
            Pis the current market price of the stock
            g is the expected growth rate of the stock price

So, if the stock price is $30.00 and the current dividend is $0.30 per quarter ($1.20 annually) and the required or expected growth rate is 8.0%, then the expected percent return is 12.0%. The required or expected return rate is determined by the market and ones own expectations for the stock, the industry, or the market adjusted for the risk of each criteria.

Therefore, if a company will be examining various business investments such as plant, equipment, or property for a future plant or other use, then it must know its cost of capital to use as a base. This is necessary to further outline what is an acceptable return for its future investments. **Remember, the goal is to enhance stockholder value.** Therefore, why would one invest into a project that returns less the cost of money adjusted for risk. How much is the cost of money? See the table below.

XYZ Corporation Cost of Capital						
	a	b	c	d	e	f
Capitalization	Units	Price	Market Value	%Total	Market Rate of Return (or cost to the company)	Extension
Bonds@ 7.0%	50,000	$1,000.00	$50,000,000	14.7%	4.2%	0.0617
Preferred Stock	500,000	$ 100.00	$50,000,000	14.7%	9.0%	0.0132
Common Stock	8,000,000	$30.00	$240,000,000	70.6%	12.0%	0.0847
TOTAL			$340,000,000	100.0%		0.1041
The Cost of Capital for the XYZ Corporation is:						10.4%

Bond cost @4.2% is the after-tax cost to the company (7.0% pre-tax and a 40% income tax rate).

This table is a weighted-average calculation. Per the above discussion, the number of units of each outstanding security is multiplied by its market price. This results in the value of the securities held by the public, the Market Value (a*b=c). Column d calculates the distribution of the value of each type to the total. Per the above discussion, column e details the rate required by the market for each. Column f extends the calculation (d*e, the weighting percentage by the rate for each, respectively) and then sums it up to the weighted average.

Business Investment A nalysis

Project Investment Return

Now that we know that the cost of capital for the company is 10.4%, we can use this to generate our requirements for investment, which is to increase shareholder value.

Project Categories

Not all projects are alike! There are several types of projects. Each type has its own requirements and risk!

1. **Increase the working capital**: This type of investment is for general operations (sometimes for inventory), or it can be for a variety of reasons. This is usually associated with increased utilization of a plant or plants to be able to provide the same product to more customers. This is low risk since; if it fails (and if the products are not perishable), the added inventory can be sold to existing customers. It probably would take an investment return equaling the cost of capital. Some additional percentage points should be added for the small risk and to earn a profit on the investment. The return is the added marginal profits. Required rate of return could be 12.0% - 15.0% in the environment under discussion.

2. **Cost reduction investment**: This type of investment is also relatively safe. It is generally for equipment to improve the operation of the facility, usually new and improved technology of all sorts. The equipment is usually already proven, and if it fails the investment is lost, but not anything more as the former "technology" may be reinstalled. Cost reduction or product improvement is necessary to maintain a competitive edge (to keep up with competition, to constantly keep the product fresh, and to extend its life cycle, etc.). Also, cost reduction is low risk, but riskier than working capital. The company wants some extra return for its investment: Required Return could be 15.0%-17%.

3. **Expansion of existing product line**: The product, operations, and technology are known. We are investing in a new facility in another part of the country, or simply adding an additional production line, etc. We know the market for the products, but have never been in this geographic location before, or on a limited bases. This new facility could be a major commitment; the volume must be increased substantially to provide adequate volume for the plant to offset the net of any potential reduced volume at any other existing facility. This type of investment requires a higher return. The required rate of return could be 20.0%.

4. **A brand new product and/or facility for it, or an acquisition**: Significant risk, no experience with the market, product, etc. Required rate of return could be 25%.

There are probably more types. It's dependent upon the industry, company, management, etc. However, we have discovered the cost of money and potentially what is required for each type of project.

Why not just invest into the highest yielding projects? Most companies balance maintaining plant and equipment with core product viability, and with measured expansion. Most companies will not chance investing in all high risk investments.It is just not prudent management. Wall Street is concerned with who is managing the company and how a company's management operates.

Defining Investment Criteria: Payback Period; Rate Of Return The Time Value Of Money

If a person puts their money in a bank under the current economic environment, a bank may offer a 3.5% interest rate annually. If a person invests in various U. S. Government Treasury Securities such as Bonds, Bills, or Notes the interest rates may be 3%, or the investment may be into various corporate high quality bonds, which may yield at a rate of 7.0%-9.0%, etc. When a company invests into a project, it also wants a return on its investment. Two items were discussed above, what money costs and the rate of return criteria for the types of projects for which they would invest. Three investment-measuring criteria will be discussed here. The first is the payback period. The next two, Net Present Value and Discounted Cash Flow are very closely related and require an understanding of how the time value of money impacts the rate of return.

Payback

This is the amount of time required for the investment to be recouped. It is the pure generation of cash by the project versus the outlay of cash. It does not take into account any interest rate or valuation of the inflowing cash stream, just the absolute amount of cash received from the project. Two project examples are offered below for review and understanding. Both contain this criterion.

In the Cost Reduction Project (#1 in the Business Investment Analysis section of this chapter), the company spends $100,000 to purchase and install equipment to enhance the production of the product in year 0 (the initiation of the project). Via the annual incremental earnings and cash flow statements forecasted for the five-year analysis, the cash generated for the period forecasted are:

Time	Amount	Cumulative	Investment Payback Point
Year1	$27,950	$27,950	
Year2	27,950	55,950	
Year3	36,500	92,400	
Year4	36,500	128,900	$100,000
Year5	36,500		

(In the Investment Analysis of this chapter [the financial analysis of a cost reduction project] the detail development of these cash flows are presented):

In year 4, the last $7,600 of cash was generated in 2/10th of the year ($7,600/$36,500). Thus, the cash investment for this project was recouped in 3.2 years. The concept is the same as it is for all projects. For this Cost Reduction project, recovery in 3.2 years is acceptable relative to the goals of the company. The drawback for this criterion is if this were all that were analyzed, the analysis and consideration would stop there. What happens after that point would not be known. It could crash completely, but would not be apparent.

We could also create a similar Cost Reduction project in which the same $100,000 is invested. However, no cash return is generated until year 4 when $130,000 is generated in February. Again, the payback is 3.2 years, but there is no interim of cash flow and thus, a greater risk. We will revisit this $130,000 example at the end of the Net Present Value (NPV) discussion.

The Time Value Of Money And Projects Based Upon It

Let's start with the basics! You put money $100.00 into a savings account at 5%. At the end of the first year, you have $105.00. Let it ride, again it draws interest at 5%. At the end of the second year, you have $110.25 ($105.00 x 1.05, five percent on top of the base 1.00), again you let it ride, $110.25 x 1.05 = $115.76. etc. Therefore, if you invest $100.00 at 5% per annum, it compounds to $115.76 at the end of the third year.

Now let's take the opposite, you could receive $100.00 today or $100.00 in three years. Choose one. Obviously, take the "now" receipt.

One other, you could receive $86.40 now or $100.00 in three years, and if 5% were guaranteed with an absolutely safe company, which one would you choose? The answer is complete indifference. This is because, if the $86.40 were invested at 5% per annum, compounded, at the end of three years it would be worth $100.00. Thus, $100.00 three years from now at a 5% discount rate is worth $86.40 in today's money. The formula for compounding is:

Future Value = Present Value $ x (1+i) n

The formula for discounting is the opposite:

Current Value = $/(1+i)n

Where $ is the cash flow per annum
i is the interest rate
n is the number of years

Again, this is for a one year, one particular year's cash flow. For a stream of cash flows, the formula is:

Current Value = (r) $Y1/(1+i)1+$Y2/(1+i)2 + $Y3/(1+i)3, etc

From a practical standpoint, the "factor" for each year is: 1/(1+i), that divided by (1+i) to generate the factor in year 2 and that divided by (1+i) to derive the factor for year three, etc.

So, if the company requires a 15% return, then the factors for the three years are:

Year	Calculation	Factor
Year 0	1.000	
Year 1	1.00/1.15	.870
Year 2	.870/1.15	.757
Year 3	757/1.15	.658, etc.

Check the Cost Reduction Project. The factors are these and can be derived by checking the table in the Appendix for discount factors. A Compounding Factors Table is also provided and is labeled as such.

Reiterating, essentially the table shows that cash received in the future is not worth as much as it is today, and given an interest rate that value can be calculated in today's dollars. The higher the interest, the lower the factors are in each future year, essentially reducing the present value of the cash flow in that year.

Net Present Value

This is an analytical method wherein all incremental activities (e.g. increased sales or improved efficiencies and less additional operating costs) generate incremental annual cash flows. These future cash flows are discounted at a desired or required rate of interest (representing the required interest rate by the company or investor). If the Present Worth (PW) of the discounted future cash flows is greater than the investment, it is acceptable.

Mechanically, a stream of annual cash flows is calculated via an incremental earning statement for the project usually for five years (although ten years has also been used, depending upon the project and the culture of the company). Each of the annual cash flows is discounted by that particular year's discount factor for the rate chosen. Then those values are summed and the initial investment is subtracted. If the result is positive, then the sum of the future values discounted at the required rate is in excess of the investment. It, thus, exceeds the required return. In the Cost Reduction example, the 15% discount factor generated a Present Worth of the future cash flow of $108,454, less the $100,000 investment, which passed the return minimum requirement. For the Expansion Project, the required return was 20%, the discounted flows totaled $29,093 in excess of the $700,000 investment. Again, it qualified for further consideration.

See the financial analyses made for the projects in the Business Investment Analysis section of this chapter below.

For a moment, let's revisit the "payback" example, previously discussed. $100,000 was invested and 3.2 years later $130,000 was returned!

Analyzing this flow of funds and incorporating the time value of money via the Net Present Value method, an unsatisfactory rate of return would be generated of about 5.3 %. For it to return an NPV of 15% the $130,000 would have to grow to $201,000 (in February of Y4). (One additional point, in periods of high interest, the annual cash flows generated could also undergo discounting to create a Present Worth Payback Period.) So let's talk about discounting.

Discounted Cash Flow (DCF or Internal Rate of Return, IRR)

This is nearly the same as for NPV. However, instead of reporting the excess (or shortfall) of the PW of the yearly cash flows, relative to the invested amount, it determines the interest rate that will cause the inflows to exactly match the investment outflow. That will be the perfect interest discount rate. And, again, obviously, the higher the rate the better as it has to "knockdown" or reduce the value of future cash flows to a greater extent in order to match them. In the Cost reduction project, at 15%, there was a net present value of $8,454. Testing the discounting at 18% generated a positive $756 (total PW is $100,756). Close but no cigar! At 19%, there was a shortfall of $1,619, as the total PW was $98,381. Thus, we know the exact rate is somewhere between 18% and 19%.

Interpolation (determining the exact point between the two rates) says: since the difference totals $2,375: $756 is 32% of the difference; up from the 18%, the higher value. Thus, 18.3% is the exact point between the two rates. Therefore, per our best forecast, the project yields at 18% versus the required "hurdle rate" of 15%. It should be undertaken or included in the list of potential satisfactory projects.

While the 18.3% is "accurate," considering that this was developed from future estimates, splitting hairs this close is a bit much. 18% should be sufficient. A project should not have to hang on .3%.

A financial calculator does the exact same thing, but faster. A computer spreadsheet also does the exact same thing even faster. The equipment all uses the same trial and error method as the manual method calculates—it just does it faster.

Investment Analysis

All the techniques will answer two questions. FIRST QUESTION: What is the net incremental benefit versus the investment? SECOND QUESTION: What is the return when the timing of the benefits is considered? It must be noted that the investment analysis is always measured on a cash basis. Cash OUT versus Cash IN and the time required to generate the inflows. Project Cash Flow!

INVESTMENT DEFINED: From a pure accounting standpoint, the investment consists of the cost of the equipment plus "capitalized expenses." Capitalized expenses are what it costs to get the equipment in and operating: the purchase of the equipment (including all taxes), the cost of delivery, the cost of installation (installation labor, materials, and services), including internal labor, if that is used. From an accounting standpoint, that is what is entered into the asset accounts for Plant, Equipment, etc. The cost of testing is not included. Those items are usually made a part of the first year operating expenses. All the above is part of the cash investment, and is depreciated based upon classification. Land is never depreciable.

This paragraph addresses the FIRST QUESTION, what is the net incremental benefit from this project versus the investment and how is it determined? The project analysis is detailed, year by year, in an incremental earnings statement for the project. The investment is made (in theory) during year zero, and the benefits are reaped in the following years. Anything more complex than that should go to a financial analyst experienced in this and its intricacies. The benefits must be incremental, i.e. added to the company earnings and cash as a result of the investment. No allocations are here of any sort. If it is a cost reduction project, then what costs will be reduced (e.g. labor, plus payroll expenses and fringe benefits, reduced material usage or material substitution, etc.)? There may also be additional operating costs and depreciation on the new equipment investment. If it is an expansion project, then what are the additional revenues, less the additional costs, and expenses. These are items that would not be there if the investment were not made. For example, if there are added sales, then probably, there are additional costs to be deducted; such as additional direct costs and additional incremental plant costs, or new added plant cost if the expansion is at an existing facility. Again, what is the net incremental impact?

Investment Analysis; the Shorter Method

Many investment opportunities (such as equipment purchase for cost reduction or expansion) can be analyzed using the principles discussed within this chapter, but via a simpler method than is presented further in this chapter.

In the first example, equipment is purchased to enhance the efficiency of the production operation. It proposes savings in both materials and labor. OK, so if that is all that will happen, it would lend itself to a less complex analysis, but, reiterating, incorporating the basic financial principles discussed.

In this analysis, volume is assumed to be stable throughout the period of analysis (5 years for either tax, engineering life or practical use with no salvage value at the end of the project).

Investment: $100,000	
Savings: Labor:	
Materials:	$5,250
	28,000
Less:	
Depreciation	
Annual:	(20,000)
Change in Pre-tax Earnings	$13,250
Income Taxes @ 40%	5,300
Incremental Net Profit	$7,950
Add back non- cash expense	
Depreciation	20,000
Annual Cash Net Income	$27,950

Factor (which also equals the payback definition)= 100,000 / 27,950 = 3.58 To find the Rate of Return, go to Appendix A-3 and along the Year 5 line locate the closest factor to 3.58. It is 13% (not quite meeting the hurdle rate). However, with accelerated depreciation for taxes and most likely some volume increase in future years, it should make the required 15% rate and be worthwhile. Also note, if your effective tax rate is below the 40% used in this example, then the INCREMENTAL NET INCOME and the

Cash Net Income will increase; and so will the rate of return. See the more detailed analysis below.

As a point of information, Appendix A-3 is the sum by year of the factors in appendix A-1, the basic discount table; and is used for level cash flows during a period of several years to ease the computation. To check this out, add the factors in the 18% column in appendix A for 5 years (.847, .718, etc) and see that it totals the sum factor in A-3.

Should an opportunity arise to expand your business via any number of means, continuing, is the investment worthwhile?

The same principles apply to the expansion as they do to a cost reduction project: What will the earnings of the company be as compared to the current?

The underlying assumption is whether the project will produce CASH to make the investment worthwhile.

Thus, with the investment into a new specialized piece of equipment:

Investment: $85,000

Incremental annual sales (@80 additional "loads" per year@ $735.00 per load

Incremental sales	$58,800
Incremental Costs	
Labor @$143 per load average	11,440
(including all fringes)	
Additional Fuel	10,000
Additional Depreciation; 7 yrs	12.000
Incremental Pre-tax Profit	$25,360
(25% tax rate)	
Incremental Net profit	19,020
Add back Depreciation (NON-CASH EXPENSE)	12,000
Cash Net Income	**$31,020**

Investment Analysis: $85,000 / 31,020 = 2.74 (payback years) Table A-3 Factor **Rate of Return:** 7 years; 2.74 = 33% (off the chart; an estimate)

If one wants to get a 20% - 25% return on their money in an expansion mode; this will be well in excess of the required return, thus worthwhile.

Further, if the cost of financing is 8%, then this states that the project is earnings 33% while costing 8%. (The cost of money is one factor in the required 20% or 25% Rate of Return calculation.)

One important note: The investment analysis determines whether the project is worthwhile. The means of financing the project, own cash or borrowing, or leasing is the next decision. See chapter 16.

Investment Analysis: The more detailed method

Below is an example of a more detailed analysis of the cost reduction project. Verbally, your company wants to invest $100,000 into cost reduction equipment. The chart created is an Incremental Income Statement for the investment of $100,000 into a cost reduction asset. It details the initial forecast for the products and the pertinent items of the cost estimated before (experienced based) and after the investment takes place. Sales and units to be sold will not change. Thus, we can assume that there will be no change in selling expenses and no change in G & A expenses as it is unreasonable to expect it and if there were, the impact would be noted. What does change is the reduction in labor as a result of the increase in machine speed (directly proportional) and the reduction in materials required to manufacture the product. This data comes from the company's production engineers. Thus, what is the impact of the new equipment? Reduced costs. Are there any increases in cost? YES, the increase in depreciation expense, which reduces the annual Pre-tax income of the company. Equipment tax life is five years and is not expected to be of any value at the end of the period. No other financial impacts are forecasted (no change in indirect labor, no change in maintenance costs, etc.) and straight-line depreciation will be used. When we take the annual statements through Cash Net Income (per Chapter Four), the result is the increase in cash annually from the project. This is what is measured against the cash investment to determine desirability.

The financial analysis details the impact of the above and culminates in three analytical criteria: Payback, Net Present Value at the required rate of return for this type of project, and the actual Internal Rate of Return (the Discounted Cash Flow, DCF). The analytical criteria answer the SECOND QUESTION (the project's rate of return) as was discussed previously in this section.

| | **XYZ Corporation** | | | | | | |
| | **Cost Reduction** | | | | | | |
	Financial Analysis						
Investment (Year 0)	$100,000						
Equipment Life: 5 years							
Annual Straight Line depreciation (no salvage value at end of life):		$20,000					
	Data Detail	Year 0	Year 1	Year 2	Year 3	Year 4	Year 5
Unit Sales	$12.00		70,000	70,000	100,000	100,000	100,000
Net Sales (basis)			$840,000	$840,000	$1,200,000	$1,200,000	
Current Operations							
Labor	$0.50		35,000	35,000	50,000	50,000	50,000
Materials	$4.00		280,000	280,000	400,000	400,000	400,000
Change in sales			0	0	0	0	0
Change in labor	15%		5,250	5,250	7,500	7,500	7,500
Change in materials	10%		28,000	28,000	40,000	40,000	40,000
Total Savings			33,250	33,250	47,500	47,500	47,500
Costs							
Depreciation			20,000	20,000	20,000	20,000	20,000
Change in Pre tax income			13,250	13,250	27,500	27,500	27,500
Income taxes	40%		5,300	5,300	11,000	11,000	11,000
Net income change			7,950	7,950	16,500	16,500	16,500
Add back depreciation			20,000	20,000	20,000	20,000	20,000
Cash net income	(100,000)		27,950	27,950	36,500	36,500	36,500
Cumulatlve Cash Flow			27,950	55,900	92,400	128,900	165,400
Net Cash (undiscounted) in excess of $100,000 investment:						21,300	86,700
Payback: years			**3.2**				
NPV@ 15%							
factor			0.870	0.756	0.658	0.572	0.497
Present Worth of <u>Cash Flows @15%</u> (Cash Net Income) Disc. Rate			24,304	21,134	23,999	20,869	18,147
					TOTAL	15% PW	$108,454
PW of C/F @18%		Factor	0.847	0.718	0.609	0.516	0.437
		Amount	23,686	20,073	22,215	18,826	15,954
					TOTAL	18%PW	$100,756
PW of C/F @19%		Factor	0.840	0.706	0.593	0.499	0.419
		Amount	23,487	19,737	21,660	18,201	15,295
					TOTAL	19%PW	$ 98,381
Disc. Cash Flow			**18.3%**				

Next is an example of an Expansion Project. Verbally, your company wants to invest into a new plant facility, which will produce the same product you are currently marketing. It is in a new location where the Marketing Department believes a good market for the product will be. Competitive pressures are such that the pricing will remain at the same levels and there will be no impact on the other facility currently producing and shipping this product. The other facility is at practical capacity and could not satisfy this new market. Thus, all the sales are additional and incremental to the company. Production engineering believes, at the least, the cost of labor and materials will be as is currently experienced and the overhead for the new facility will be at a minimum. The Production Department has prepared a budget, subject to review by the finance department. Straight-line depreciation will be used for the new facility. There is an existing sales staff in this new location, but they are not selling this product, which will be an addition to their customers. Thus, only the commission will be incremental, 10% of net sales value. To analyze this project, an incremental earnings statement is produced for the 5 years of the forecast, Sales through Net Earnings through Cash Net Income. As this project is an expansion project, additional Working Capital must be invested: inventory, accounts receivable on the additional sales, and (less) additional accounts payable on the materials used. See the schedule at the bottom. Also note, at the end of the five-year life of the analysis, the working capital balance at the end is recovered. The three main analytical criteria are presented.

	Investment		!Years Tax Life	Annual Depreciation
		XYZ Corporation		
		Expansion Project		
		Financial Analvsis		
Equipment		$500,000	5	100,000
Building		200,000	30	6,667
Total		700,000		106,667

		Year 0	Year 1	Year 2	Year 3	Year 4	Year 5
Units Forecasted			65,000	80,000	100,000	100,000	100,000
Sales Price F/C	$12.00		$780,000	$960,000	$1,200,000	$1,200,000	$1,200,000
Costs of Goods Sold							
Dir Labor	$0.50		32,500	40,000	50,000	50,000	50,000
Dir Materials	$4.00		260,000	320,000	400,000	400,000	400,000
New Plant Expense			50,000	50,001	50,002	50,003	50,004
Depreciation			106,667	106,667	106,667	106,667	106,667
TOTAL Incremental COGS			449,167	516,668	606,669	606,670	606,671
Incremental Gross Profit			330,833	443,332	593,331	593,330	593,329
%Sales			42.4%	46.2%	49.4%	49.4%	49.4%
Increm Selling Exp.	10.0%		78,000	96,000	1200,000	120,000	120,000
Increm Pre-tax			252,833	347,332	473,331	473,330	473,329
Income Taxes	40.0%		101,133	138,933	189,333	189,333	189,332
Increm Net Earn			151,700	208,399	283,999	283,998	283,998
%Sales			19.4%	21.7%	23.7%	23.7%	23.7%
Add Back Depreciation			106,667	106,667	106,667	106,667	106,667
Less: Add'l W/C			(113,264)	(26,138)	(34,850)	0	174,252
Cash Net Income		(700,000)	145,103	288,928	355,815	390,665	564,916
Cum C/F from Ops			145,103	434,031	789,846	1,180,511	1,745,427
Net Cash (undiscounted) in excess of $700,000					89,846	480,511	1,045,427
Payback: years			2.9				
NPV@20%							
	Factor	1.000	0.833	0.694	0.579	0.482	0.402
*	Amount	(700,000)	129,919	200,645	205,911	188,399	227,027
Present Worth of Net Discounted Cash Flows @20 %							242,902
*on annual Cash Net Income							
Discounted Cash Flow (IRR)			32.0%				
Working Capital Detail							
Increm Inventory 30 days			37,431				
Increm A/R; 45 days			97,500				
Increm A/P 30 days			(21,667)				
Net Increm W/C			113,264	139,402	174,252	174,252	174,252
additional annual W/C			113,264	26,138	34,850	0	0

Analytical Criteria are the same as previous. A) Payback period of 3.0 years ($502 is not material). That is the point where the pure raw cash in, not discounted, equals the investment outflow of cash. B) **The Net Present Value** of the cash flow totals a positive $29,093. This is the discounted value at a 20% discount rate of the future cash inflow ($729.093 less the present worth of the outflow, $700,000 at a discount factor of 1.000; since all is spent in that year). Thus, at 20% the present value is positive, indicating that it is at least in excess of 20%. C) The DCF is 21.7% (also above the required 20% "hurdle rate").

The rates of return for these projects, the DCF rates, were not arbitrary. In the preceding section on Cost of Capital, our company set up minimum rates for different types of investment projects. If these "hurdle rates" are part of company policy, then the projects submitted for consideration must meet the rates or not be considered, period!

Capital Budgeting

Essentially, this involves the selection of the projects to be undertaken, a process of "ranking the projects." Several quantitative aspects are examined. How much cash does the company have now and in the foreseeable future from internally generated and via borrowing? This is where the detailed financial plan and longer-range plan are brought into play regarding how much cash will be generated with and without the projects, and the impact of each. How much does the company want to finance, via borrowing (there is an impact on the debt-equity ratio and times interest earned ratios)? This also impacts the interest rates and thus the cost of capital. And let's not forget the Sustainable Grow Rate. Does the company want to finance the project with new equity financing? What is the impact on Earnings Per Share before and after the impact of the projects? After all, the goal is to enhance shareholder value.

Several capital budgeting methods are employed. One is to simply separate all the projects into categories or types and rank by DCF. It may be the company's policy to assign an importance indicator; e.g., "A" must do; "B" should do; "C" can do, if we have an excess of funds. Then, more qualitative analysis is applied regarding the markets for the products and the life cycles, etc.

Thereafter, again lay out the spending for those projects selected as an integral part of the Cash Flow of the three-year or five-year Finance section of the Business Plan.

That's it! END Capital Budgeting!

Addendum

Bond Price Analysis

To further demonstrate the impact of interest rates, this discussion will outline the valuing of bond pricing. It is designed to show why and how this type of security is priced and what happens when market conditions change. It is also designed to reinforce the Net Present Value analysis.

See the example below!

Example: A seller is asking a buyer to purchase a piece of paper, which "guarantees" to pay the purchaser $70.00 per year for 10 years and at the end of 10 years to also return the purchaser's original $1,000. In short, it is a bond. If the market conditions for this type of security are 7%, then using the Present Worth method and factors, the present value of all those cash flows is $1,000.

If the security is locked in at paying $70.00 per annum, but the market interest requirement has declined to 6%, then the value of the future $70.00 interest payments and the $1,000. "redemption" at the end of the period is worth more.

Thus, the value and price of the bond will rise, in this case to $1,073.60. The holder has a potential appreciation of value if it is sold at a favorable moment. However, if held to maturity, only the $1,000.00 face value will be paid. The opposite is true if the market interest rates move up.

XVZ Company
10 year 7% Debenture Bond
Face Value $1,000. Per unit

Market interest rate for this type and risk type ofsecurity is 7%.
Annual dividend $70.00

PW factor@ 7%

	Year 0	Year1	Year2	Year3	Year4	Year5	Year6	Year7	Year8	Year9	Year10	Total PW
discount factor	1.000	0.935	0.873	0.816	0.763	0.713	0.666	0.623	0.582	0.544	0.508	
Dividends		$ 70	$ 70	$ 70	$ 70	$ 70	$ 70	$ 70	$ 70	$ 70	$ 70	
Redemption of Principal											$ 1,000	
Present Worth		$ 65.42	$ 61.14	$ 57.14	$ 53.40	$ 49.91	$ 46.64	$43.59	$ 40.74	$ 38.08	$35.58	$ 491.65
											$508.35	$ 508.35
											Total Present Worth	$1,000.00

"What if" the market rate for this type and risk type of security declined to 6%

PW factor@ 6%

	Year 0	Year1	Year2	Year3	Year4	Year5	Year6	Year7	Year8	Year9	Year10	Total PW
	1.000	0.943	0.890	0.840	0.792	0.747	0.705	0.665	0.627	0.592	0.558	
Dividends		$ 70.00	$ 70.00	$ 70.00	$ 70.00	$ 70.00	$ 70.00	$ 70.00	$ 70.00	$ 70.00	$ 70.00	
Redemption of Principal											$1,000.00	
Present Worth		$ 66.04	$ 62.30	$ 58.77	$ 55.45	$ 52.31	$ 49.35	$ 46.55	$ 43.92	$ 41.43	$ 39.09	$ 515.21
											$ 558.39	$ 558.39
											Total Present Worth	$1,073.60

CHAPTER SIXTEEN —
LEASE VERSUS PURCHASE

A lease is a means of obtaining a piece of equipment. It is a financing instrument. The equipment can be acquired by using the company's own funds or the company may choose to borrow the funds and purchase the equipment, or the company may enter a lease contract to "use" the equipment. The lease allows the company to use the equipment without any initial cash expenditure and without the necessity of borrowing from a financial institution. There are two principal types of leases: an operating lease and a financial lease.

Operating Lease

The Operating Lease is the simplest. It is a contract wherein the lessee has use of the equipment for a period of time with the option of canceling upon short notice. The total of the payments will generally not exceed the equipment's value. Transfer of ownership is not contained within the lease. This type of lease is generally utilized for equipment too expensive to purchase as it is not needed for an extended period of time. The equipment may only be needed for one particular requirement that when completed is no longer necessary. Or technology is such that the equipment to be leased will be obsolete very quickly as far as its use in your operation is concerned. Whatever the reason, the company acquires the piece of equipment for a specified period of time or on a monthly basis and pays for it accordingly. The lease contract may or may not include maintenance, insurance, pickup, and return, etc. The lease is simply accounted for as an operating expense and placed in the proper grouping of accounts (cost of sales; general and administrative expenses, etc.)

Financial Lease

A Financial Lease is a long-term contract and under which, the lessee (your company) formally agrees to make a series of payments over a period of time. The total of the payments is in excess of the asset's purchase price. The lease payments are usually spread out over a period of time, but not in excess of the life of the asset. This type of lease must be accounted for as a capital purchase (i.e., placed on the books as an asset and depreciated). The payment is the lease obligation (set up as a long-term liability) and reduces via the accounts payable system. This type of lease MUST be capitalized if it meets at least one of the following criteria:

1. The lease transfers the ownership of the asset at its end.
2. The lease contains an option to buy at an unusually low price.
3. The term of the lease is at least 75% of the estimated life of the asset.
4. The present worth of the minimum lease payments is at least 90% of the asset's fair market value.

Note: The fair market value is usually the price one would reasonably expect to pay for it in the open market. The present worth factor to use is the company's long-term borrowing rate. Speak to your bank in this regard; it's probably around 10%.

There are several analyses that are made. The initial analysis centers around whether the acquisition of the asset is worthwhile. This analysis is performed as was discussed in Chapter 15, Investment Analysis. The equipment may be utilized for cost reduction purposes or for expansion of sales (additional production capacity or new product). Should the investment be worthwhile, then the means of financing the equipment is the next analysis. In most financial leases, the cost of maintenance and insurance is placed upon the lessee. Therefore, these and other equipment operating costs should be made part of the initial analysis as the costs will be the same in all financing situations (purchase or lease).

Note: If the reader refers back to the Cost Reduction investment analysis presented in Chapter 15, no additional costs for maintenance, etc. had been included as it was a replacement and similar costs were already being incurred with no material changes estimated for this new piece of equipment.

The financing analysis would look like this.

First, what would a lease payment schedule look like and how is it calculated? Below is the manual calculation for a loan repayment, unamortized and then amortized (it creates the level payments). The small difference is due to the timing of the repayment. A financial calculator or a computer spread sheet program will do the same, but this is for demonstration purposes.

XYC Company Five-Year Loan Repayment Schedule

A $100,000 lease with a S year payment at 10% interest

Unamortized

	Opening Balance	10% Interest	Principal	Total Payment	Year-End Balance
Year 1	$100,000	10,000	20,000	30,000	80,000
Year 2	80,000	8,000	20,000	28,000	60,000
Year 3	60,000	6,000	20,000	26,000	40,000
Year 4	40,000	4,000	20,000	24,000	20,000
Year 5	20,000	2,000	20,000	22,000	0
Totals		30,000	100,000	130,000	

Annual Average Payments: $ 26,000

Amortized

	Opening Balance	10% Interest	Principal	Total Payment	Year-End Balance
Year 1	100,000	10,000	16,380	26,380	83,620
Year 2	83,620	8,362	18,018	26,380	65,602
Year 3	65,602	6,560	19,820	26,380	45,782
Year 4	45,782	4,578	21,802	26,380	23,980
Year 5	23,980	2,398	23,980	26,380	0
		31,898	100,000	131,900	

Thus, for a "loan" of any sort, with a 10% interest rate (even for a lease), the actual interest calculations would look as above. However, with a bank loan, the schedule would be printed and provided to the borrower. With a lease carrying the same rate (unless the lessee were so advised), all the lessee would find out would be the annual or monthly payments.

However, as the below analyses reveal, it will be used as is necessary.

In the first analyses, what is compared is the use of a company's own funds versus borrowing them from a bank (at the 10% interest rate used in the above example.) Don't confuse this with the Present Worth **factors** used to calculate the Present Worth of the cash flows to determine whether the project is worthwhile (meeting the "return" requirement or "hurdle rate").

Repeating, the XYZ Corporation will purchase a new replacement machine. The analysis shows that at a 10% cost of money to XYZ, it is economically desirable for the company to use its own funds. Please, note that straight-line depreciation has been replaced with MACRS rate depreciation on a tax basis as it is always more advantageous. It is most likely that taxes will be calculated using this depreciation schedule. (Although, it is also likely, and legal, to carry the property on the books on a straight-line depreciation basis.) One additional point, these types of leases usually contain a minimal buyout provision (sometimes $1.00 or $100.00 at the end of the lease and this transfers the ownership to XYZ Corporation. Thus, this is not a consideration.)

<table>
<tr><td colspan="10" align="center">XYZ Corporation
Lease Vs. Purchase Analysis</td></tr>
<tr><td colspan="10" align="center">Own Funds vs. Bo Funds</td></tr>
<tr><td colspan="5">Purchase</td><td colspan="5">Loan</td></tr>
<tr><td colspan="2">Investment $100,000</td><td>10%</td><td></td><td></td><td></td><td></td><td></td><td>10%</td><td></td></tr>
<tr><td></td><td>MACRS Depreciation</td><td>Depreciation Tax Savings</td><td>P. W. Factor</td><td>Present Worth</td><td>Payment</td><td>After tax Interest</td><td>Net Cash Payment</td><td>P. W. Factor</td><td>Present Worth</td></tr>
<tr><td>Year 1</td><td>20,000</td><td>8,000</td><td>0.909</td><td>7,273</td><td>26380</td><td>6,000</td><td>20,380</td><td>0.909</td><td>18,527</td></tr>
<tr><td>Year 2</td><td>32,000</td><td>12,800</td><td>0.826</td><td>10,579</td><td>26380</td><td>5,017</td><td>21,363</td><td>0.826</td><td>17,655</td></tr>
<tr><td>Year 3</td><td>19,000</td><td>7,600</td><td>0.751</td><td>5,710</td><td>26380</td><td>3,936</td><td>22,444</td><td>0.751</td><td>16,862</td></tr>
<tr><td>Year 4</td><td>12,000</td><td>4,800</td><td>0.683</td><td>3,278</td><td>26380</td><td>2,747</td><td>23,633</td><td>0.683</td><td>16,142</td></tr>
<tr><td>Year 5</td><td>12,000</td><td>4,800</td><td>0.621</td><td>2,980</td><td>26380</td><td>1,439</td><td>24,941</td><td>0.621</td><td>15,487</td></tr>
<tr><td>Year 6</td><td>5,000</td><td>2,000</td><td>0.564</td><td>1,129</td><td>26380</td><td>6,000</td><td>20,380</td><td>0.909</td><td>18,527</td></tr>
<tr><td></td><td>100,000</td><td></td><td></td><td></td><td></td><td></td><td></td><td></td><td></td></tr>
<tr><td colspan="4">Total P W of Depreciation Tax Savings:</td><td>$30,949</td><td></td><td></td><td></td><td></td><td>$84,673</td></tr>
<tr><td colspan="4">Net Cost: Investment (current) less</td><td>100,000</td><td></td><td></td><td></td><td></td><td></td></tr>
<tr><td colspan="4">PW of Depreciation Tax Savings</td><td>30,949</td><td></td><td></td><td></td><td></td><td></td></tr>
<tr><td colspan="4">Net Cost</td><td>$69,051</td><td></td><td></td><td></td><td></td><td>$84,673</td></tr>
</table>

* Tax Rate is 40%

* MACRS allows a one-half year depreciation in the first year.

However, if a company has a cost of money of 15% or it requires such a return (as in a "safe" cost reduction investment), then the analysis would look like this.

<table>
<tr><td colspan="10" align="center">XYZ Corporation
Purchase Analysis
Own Funds vs. a.... Funds</td></tr>
<tr><td colspan="5">Purchase</td><td colspan="5">Loan</td></tr>
<tr><td colspan="2">Investment $100,000</td><td>10%</td><td></td><td></td><td></td><td></td><td></td><td>10%</td><td></td></tr>
<tr><td></td><td>MACRS Depreciation</td><td>Depreciation Tax Savings</td><td>P. W. Factor</td><td>Present Worth</td><td>Payment</td><td>After Tax Interest</td><td>Net Cash Payment</td><td>P.W. Factor</td><td>Present Worth</td></tr>
<tr><td>Year1</td><td>20,000</td><td>8,000</td><td>0.870</td><td>6,957</td><td>26380</td><td>6,000</td><td>20,380</td><td>0.870</td><td>17,722</td></tr>
<tr><td>Year2</td><td>32,000</td><td>12,800</td><td>0.756</td><td>9,679</td><td>26380</td><td>5,017</td><td>21,363</td><td>0.756</td><td>16,153</td></tr>
<tr><td>Year3</td><td>19,000</td><td>7,600</td><td>0.658</td><td>4,997</td><td>26380</td><td>3,936</td><td>22,444</td><td>0.658</td><td>14,757</td></tr>
<tr><td>Year4</td><td>12,000</td><td>4,800</td><td>0.572</td><td>2,744</td><td>26380</td><td>2,747</td><td>23,633</td><td>0.572</td><td>13,512</td></tr>
<tr><td>Year5</td><td>12,000</td><td>4,800</td><td>0.497</td><td>2,386</td><td>26380</td><td>1,439</td><td>24,941</td><td>0.497</td><td>12,400</td></tr>
<tr><td>Year6</td><td>5,000</td><td>2,000</td><td>0.432</td><td>865</td><td>26380</td><td>6,000</td><td>20,380</td><td>0.870</td><td>17,722</td></tr>
<tr><td colspan="4">Total P W of Depreciation Tax Savings:</td><td>$27,628</td><td></td><td></td><td></td><td></td><td>$74,545</td></tr>
<tr><td colspan="4">Net Cost: Investment (current) less</td><td>100,000</td><td></td><td></td><td></td><td></td><td></td></tr>
<tr><td colspan="4">PW of Depreciation Tax Savings</td><td>27,628</td><td></td><td></td><td></td><td></td><td></td></tr>
<tr><td colspan="4">Net Cost</td><td>$72,372</td><td></td><td></td><td></td><td></td><td></td></tr>
</table>

* Tax Rate is 40%

Harvey Goldstein

It is still worthwhile to use internal funds, but the spread is much less as future payments are worth less via the discounting process.

Let us now examine the Lease scenario. Here is the analysis, initially at a 10% company cost of money for XYZ Corporation.

XYZ Corporation Lease Vs. Purchase Analysis									
Purchase					**Lease**				
Investment $100,000		10%						10%	
	MACRS Depreciation	Depreciation Tax Savings	P. W. Factor	Present Worth	Payment	After Tax Interest	Net Cash Payment	P.W. Factor	Present Worth
Year1	20,000	8,000	0.909	7,273	26380	8,000	18,380	0.909	16,709
Year2	32,000	12,800	0.826	10,579	26380	12,800	13,580	0.826	11,223
Year3	19,000	7,600	0.751	5,710	26380	7,600	18,780	0.751	14,110
Year4	12,000	4,800	0.683	3,278	26380	4,800	21,580	0.683	14,739
Year5	12,000	4,800	0.621	2,980	26380	4,800	21,580	0.621	13,399
Year6	5,000	2,000	0.564	1,129		2,000	-2,000	0.564	-1,129
Total P W of Depreciation Tax Savings:				$30,949	Net PW Cost				$69,052
Net Cost: Investment (current) less				100,000					
PW of Depreciation Tax Savings				30,949					
Net Cost				$69,051					

   * Tax Rate is 40%

At a 10% cost of money or required rate of return for this financial "investment," it's a wash; but favorable to use of the company's own funds.

However, at a 15% requirement:

XYZ Corporation Lease Vs. Purchase Analysis									
Purchase					**Lease**				
Investment $100,000		15%						15%	
	MACRS Depreciation	Depreciation Tax Savings	P.W. Factor	Present Worth	Payment	After Tax Interest	Net Cash Payment	P.W. Factor	Present Worth
Year1	20,000	8,000	0.870	6,957	26380	8,000	18,380	0.870	15,983
Year2	32,000	12,800	0.756	9,679	26380	12,800	13,580	0.756	10,268
Year3	19,000	7,600	0.658	4,997	26380	7,600	18,780	0.658	12,348
Year4	12,000	4,800	0.572	2,744	26380	4,800	21,580	0.572	12,338
YearS	12,000	4,800	0.497	2,386	26380	4,800	21,580	0.497	10,729
Year6	5,000	2,000	0.432	865		2,000	-2,000	0.432	-865
Total P W of Depreciation Tax Savings:				$27,628		Net PW Cost			$60,802
Net Cost: Investment (current) less				100,000					
PW of Depreciation Tax Savings				27,628					
Net PW Cost				$72,372					

* Tax Rate is 40%

At a 15% required rate of return, it is significantly worthwhile to lease.

Under this scenario, it is better to lease than to purchase. So why don't we lease everything! Some companies lease as much as they can. However with this business philosophy, there may be significant personal guarantees required by the financing institution or the companies are very well established and profitable.

In addition, consideration will be given to the company's debt/equity ratio, its current ratio and the impact on its times fixed charges ratio. The consideration is both internal (management and ownership) and external (banks, other debt holders). If there is another debt that the company has, it is also likely that there are restrictions regarding the addition and type of debt that the company may incur beyond the debt already in place. And, there may also be restrictions on dividend distribution.

Obviously, each situation and each company is different. This chapter was designed to outline the mechanics and to discuss some of the ramifications and other items that have to be considered when deciding to lease or purchase equipment or assets.

CHAPTER SEVENTEEN —
CREDIT AND COLLECTIONS

This discussion centers on the issuance of credit to your customers. Generally, a retail business will deliver the product and receive payment for it at the point of sale. Thus, no credit is offered. However, when the sale of your product or service can be paid for at a future date, you are offering credit. You are now acting as a bank.

Issuing Credit

The fundamentals for a new account are to obtain credit references from three current suppliers and a reference from the customer's bank. This information should be secured via your company form, similar to what is asked of your company. See sample forms attached at the end of this chapter. What questions should be asked of credit references?

1. How long has the customer been doing business with the reference?
2. What is the highest credit balance this customer had?
3. What is their average credit balance?
4. What is their payment history (prompt, slow 15 days, etc.)?
5. Other comments

From the company's bank, your company would want to know:

1. How long has the company done business with the bank?
2. What has their average balance been?
3. Have their been any return items?
4. Are their any other comments that the bank wishes to make?

In addition, depending upon the number of customers for which credit will be involved, a credit reporting agency may be engaged (such as, Dunn & Bradstreet). This and other similar companies will provide their rating of the company investigated on their own scale. Also, depending upon the depth of analysis, most likely, they will be able to supply various reports differing in scope. Many industry or trade associations have similar credit services.

New customers may not be issued credit as you may request "Cash on Delivery" (COD) for the first delivery, if it is a standard product. If the initial purchase is a special product, then your company may ask for 50% in advance and the balance COD. For the special products, the upfront request is to somewhat cover costs should the customer renege on the purchase and your company is stuck with a potentially un-saleable product. You may also want to do this for customers you have done business with, and are questionable for some reason.

After a period of time, with your customer paying COD or even after the first order, your company may place the customer on "normal" terms within a certain credit limit. The LIMIT may be determined by the customer's limited history with you, a forecast or request for a high limit from them, and the customer's credit rating. All, of course, measured against the customer's payment history.

Depending upon the size of the customer and its credit requirements, your company may request periodic and annual financial statements from the customer.

Managing Credit

A company may choose to have an individual charged with overseeing the function of credit management. This is dependant upon the volume of credit issued. The individual may devote part of a workday concentrating upon this function or it may be necessary to employ a full-time credit manager. The function must be performed every day and would include a constant review of the payments and outstanding invoices for each customer. A review of the credit limit of each customer is made periodically. It would also include tracking of uncollected invoices, that periodically "fall through the cracks," even from good customers.

Many companies produce and mail **monthly statements** as a prompt for collecting the outstanding invoices. This is time-consuming and costly. However, while most companies pay from invoices, mailing monthly statements to selective companies (generally large with multiple locations or branches) can be a helpful tool for account reconciliation, for the customer and yourself. It is the style of the industry and company as to whether this should be undertaken.

One other function is the enforcement of the company's credit policy. Please refer back to Chapter Seven, Working Capital regarding the parameters and impacts of various credit policies. (Remember, this will impact sales, earnings, and cash flow.) If the company decides to issue credit with the net balance due within 30 days, when is the friendly prompt call made? Is it made on the 31st day? The 35th day? The 45th day? This will set up the pattern for the company. However, when the policy is chosen, the contacts with the customer must be made with consistency. This will ensure that the customer is aware of your policy and will help to ensure that funds are received on a regular basis.

Should invoices go past the due date and after contact(s) is made, the decision has to be made whether to stop future shipments until the unpaid invoices are collected. This will depend upon how big your business is and your knowledge of the customer. The customer knowledge impacts how much of an extension will be tolerated. It is a business decision and usually, depending upon the size of the account, may be subject to review by higher authorities.

Collecting "Past due" receivables

There are times when your customers may encounter tight cash situations. Then a business decision has to be made whether to extend the collection. If the customer promises a date when the payment is to be made and meets that commitment, then they look like a hero; if the customer does not meet it, then his reputation is usually tarnished. It is the same with your own accounts payable operation or department.

Below are offered several dialogues for making telephone calls and letters. These are suggestions only. You will note the specifics regarding promises made and dates when payment was to be expected.

Telephone contact, the fastest and easiest means of contact.

The initial call should be made per policy, e.g. on the 35th day after the invoice date. The conversation should be polite and find out whether a payment was made and not yet received. If not, then mention that the payment is past the due date and the company would like it by the following "Friday" or whatever. It is important to receive a specific commitment. Notes must be made. Many computerized accounting packages that have an A/R module, have fields for such notes for access by anyone who is eligible to see such data. This central file is the best place to keep such notes. A side calendar should be kept regarding the day the next call is to be made. Thus, when that day comes, it is then easy to check to see whether the invoice has been paid.

The second call should indicate that the promised date (state the date) has passed and the company has yet to receive any payment. "I now need to know when that invoice will be paid" or something similar. Again, wait for a specific response and commitment. If none can be secured, request that the collector be placed in contact with a supervisor. The conversation may also move to question whether there is any problem that we should know about why the payment has not been made.

The third and final call should again secure a specific date within a short but reasonable time as to when the company will receive payment. You may want to indicate that unless payment is received by that date, alternate means of collection may be required. This means turning the collection over to a collection agency.

Depending upon the size of the customer, an executive of your company, the sales manager, or a senior manager, or the president may be asked to make a "personal" contact; usually the " third" call, before any stronger moves or decisions are made. Remember, the goal is to collect the invoice and maintain the account, within acceptable terms.

During this period when the collection appears to be in trouble, the credit department may want to cease shipments until the collection is made. At that time an account review may have to be made including a credit limit review, placing the customer on COD or Cash In Advance (CIA) status, recontacting other references, etc.

After each contact, a confirming letter or fax may be sent out. The initial letter, on company letterhead, may be worded as follows:

Date _______________
Dear _______________
You are a valued customer. There are times when oversights are made. So when your payment was not received as usual, we choose to call you. This is just to confirm today's conversation that arrangements have been made regarding invoice No. ____________
in the amount of $ _________. As agreed, $ ____________ will be mailed to our office no later than (date) to settle this invoice.
Thank you for your prompt attention in the matter.
Sincerely,
Senders Name and title

After the second conversation, a letter may be sent and again may be worded as follows:

On (date), I again spoke with you (or whom ever) regarding the past due amount for our invoices Nos.   , in the amount of $ respectively. Payment had been promised by (date) and has not been received. Today, a commitment was made for a payment of $ on or before (date).

We expect the payment to be received per today's commitment. We appreciate your ensuring that the payment will be made.

Sincerely,

Name and title of sender

After the third conversation or after several left messages, a letter may be sent and again may be worded as follow:

Date, Name of Company, Address, Account No., ect.,: amount, date of original debt.

We have attempted to collect this debt on a number of occasions. As of this writing, the debt is well past due with no responsible response on your part. Unless this debt is paid by ___________, we will have no choice but to seek alternative means of collection.
Sincerely,

Controller or Higher

If you're in business, any number of collection agencies will be in contact with you to provide such a service, if necessary. One or more of these agencies may be an advantage to have for such a purpose. You may want to contact their clients to aid in choosing the right one(s) for your company.

Your agency may suggest that the claim be taken to court. They usually have done this a number of times and have the know-how to have this done. They may also be able to determine how much cash the account has in its bank. However, if they have the cash today, by the time the court date comes, it may be gone. This may impact whether the court is a good final move. Most courts will not make the collection and getting a court order for payment or leaning the assets may be time-consuming. If it does not appear to be reasonably collectible, then it is time to write off the bad debt and move on.

XYZ Company, Inc.

Street Address, city and zip code
Phone: (123) 456-7890; FAX: (123) etc.

Application for Credit

Date: _____________________

Name of Applicant: _____________________
         dba: _______________
Address: _______________________

        ____

Phone: _________________________
Fax: ______________

Federal Tax ID No. or S.S.N. _____________________________

How long in business? ___________________________

Legal Form of Business:
       Corporation ____________ Partnership _________ Individual __________

Owner(s) Name: ________________________
Address: ________________________

        ____

Bank: __________________
Name: ________________________
Address: ________________________

        ____

Checking, _________ Other _______ Account No: ________

Three Trade References:
Name: _______________________________   Contact: _________________________
Address ______________________________   Phone: ___________________________
                                            Fax: _____________________________

Name: _______________________________   Contact: _________________________
Address ______________________________   Phone: ___________________________
                                            Fax: _____________________________

Name: _______________________________   Contact: _________________________
Address ______________________________   Phone: ___________________________
                                            Fax: _____________________________

By: ___________________ Title: _____________________ Date ___________

Harvey Goldstein

XYZCompany
Street address, city and zip
Phone: (123) 456-7890; FAX: (123) etc.

Credit Reference Check
The subject company has applied for credit with the XYZ Company.
We appreciate your prompt response.

To:
Name   ____

Address   ____

          ____

Phone: ____

Fax: ____

Subject:

Name of Business   ____________________________________

City, State   ____________________________________

How long has the subject company been doing business with you? ________________

Average Balance ____________________________________

High Balance   ______________________________

Terms

Payment Experience ____________________________

Comments

CHAPTER EIGHTEEN —
VALUATION OF THE COMPANY

At one point in my business career, I sold real estate. Inevitably, I would encounter an owner who would tell me that he "knows that the house is worth $300,000" or whatever he thought it was in his mind. The owner had no real basis for his valuation, just what he thought he would like to receive. Sometimes, the owner would not settle for any suggestion that the house was worth anything less; and "would not sell it for anything less." Doing my job properly, I would then perform a valuation analysis that included the size, location, and amenities. I would compare it to recent sales of properties as similar and as close to the property as possible. Most sellers would realize the "calculated" value at that point and price the property accordingly within a reasonable range. (By the way a new or remodeled bathroom does not automatically constitute an increase in the value of the house. It may look much better, from a marketing point of view; but all houses are expected to have an operating and efficient bathroom.)

The above situation discusses how one might value a home, but, how does one evaluate a business? To say "I know its worth XXXXX" is pulling the number out of the air and is difficult to determine.

There are several methods to determine the worth of a business. The two I will discuss here are:

1. The Earnings Multiple

2. The Discounted Cash Flow (DCF) Method (Yes, just as discussed in Chapter Fifteen)

The Earnings Multiple

This is a relatively easy calculation. It is usually five times EBITDAS (Earnings Before Interest, Taxes [income taxes], Depreciation, Amortization and [owners] Salaries). As an example, let us look at the data for the year completed, which was the basis for the projection detailed in the Financial Plan presented in Chapter Ten. Below, the data is repeated and the valuation is calculated.

Business Finance Was Never This Easy

XYZ Company
Earnings Multiple Basis for Business Valuation

	Actual 2xxx	% Sales		EBITIS Restated for Valuation
sesasonality				
Net Sales	6,679,709	100.0%		6,679,709
Cost of Sales				
Direct Materials	4,669,772	69.9%		4,669,772
Direct Labor	827,992	12.4%		827,992
Plant Expenses				
Utilities	50,254	0.8%		50,254
Depreciation	25,944	0.4%	**(25,944)**	0
Rent	39,452	0.6%		39,452
Health Insurance	102,202	1.5%		102,202
Shop Supplies	20,479	0.3%		20,479
Payroll Taxes	121,715	1.8%		121,715
Total Plant Exp	360,046	5.4%		334,102
Total Cost of Sales	5,857,810	87.7%		5,831,866
Gross Profit	821,899	12.3%		847,843
%sales				
General & Administrative Expenses				
Salaries (Incl. Plt. Mgr)	218,330	3.3%	**(50,000)**	168,330
Payroll Taxes	24,503	0.4%	**(5,651)**	18,852
Commissions	80,444	1.2%		80,444
Advertising & Promotion	25,611	0.4%		25,611
Auto Expenses	16,819	0.3%		16,819
Misc Txs & Licenses	4,235	0.1%		4,235
Office Expense / Postage	22,978	0.3%		22,978
General Insurance	27,655	0.4%		27,655
Repairs & Maintenance	976	0.0%		976
Dues & Subscriptions	2,280	0.0%		2,280
Travel, Meals & Entertainment	35,843	0.5%		35,843
Telephone & Utilities	15,500	0.2%		15,500
Miscellaneous	41,000	0.6%		41,000
Total Gen & Admin	516,174	7.7%		460,523
%sales				

		Actual 2xxx	% Sales		EBITIS Restated for Valuation
Income from Operations	305,725		4.6%	387,320	
%sales					
Other Income & Deducts					
Interest Expense		(62,000)	(0)	X	
Bad Debts		56,000	0		56,000
Total Other Income / Deducts		(118,000)	(0)		56,000
Pre - tax Income		187,725	2.8%		331,320
Income Taxes	35%	65,704	1.0%		
Net Income		122,021	1.8%		
		Multiple			**5**
		Estimated Sales Value			**$1,656,600**

Verbally, there is no adjustment for sales through gross profit. However, the owner's salary is $50,000 and the associated payroll taxes are proportionally reduced. The owner's salary is considered an additional expense not necessary to run the business and another means of taking money out of the business versus a dividend. Interest is caused by a means of financing and is not associated with the performance of the company. Income tax is the responsibility of the new owners after they "merge" the company into their ownership and their tax situation. Again, income taxes are not reflective of the performance of the business; whereas, operating earnings (essentially EBIT) are. In some formulas, the bad debts charge is also eliminated as it is not reflective of the performance of the business. I have included it, as I am assuming it is reflective of the industry.

In this case, the chart calculates the theoretical sales price as $1,657,000.

Note: In this example, no value or "price" is quoted for the assets of the company being acquired. It assumes that the purchase price agreed upon will be for the total business. Anything is negotiable. However, based upon the above, one could argue that: 1) as the business yields an excellent Return on Assets (ROA) or Return on Equity (ROE), that the assets are minimal and the earnings are good, thus, the multiple reflects this (incorporating the assets into the whole business being purchased); or 2) as the ROA or ROE of the current company is poor and the assets are substantial, that the sale price is also reflective of the poor return and includes the assets.

The FIVE multiple is a general rule of thumb. It can be adjusted up or down depending upon various factors. Several could be:

- Is the company out performing the industry?
- Is the company growing or stagnant or in decline?
- Is the industry a growth industry or nearing the end of its life cycle?

Note: A slew of other factors may influence the multiple up to possibly six or seven or down to four or three or someplace in between.

The DCF Method

This is the method most commonly used by a larger company when it is evaluating an acquisition.

I was involved in the acquisition of a major new product and plant developed by one company and desired by the company where I was employed. In summary, the financial analysis determined that the acquisition would surpass the company's required Discounted Cash Flow rate of return for an acquisition to expand a product line (25%). However, it was dependent upon a new formula, which was only "weeks away." The acquisition was made, but the new formula never came to fruition and the sales forecast was not met. Lots of finger-pointing took place. Essentially, the assumptions were a bit too optimistic. The Finance Department so noted and discussed the assumptions, especially the non-proven new formulation. Yes, mistakes were made.

XYZ Company
Projection for acquisition
Annual Growth Rate   7.5%

	Actual 2xxx	% Sales	Year 1	% Sales	Year 2		Year 3		Year 4		Year 5	
Net Sales	6,679.709	100.0%	7,300,000	100.0%	7,847,500		8,436,063		9,068,767		9,748,925	
Cost of Sales												
Direct Materials	4,669,772	69.9%	4,972,698	68.1%	5,297,063	67.5%						
Direct Labor	827,992	12.4%	865,252	11.9%	902,463	11.5%						
Plant.Expenses												
Utilities	50,254	0.8%	55,279	0.8%	60,807							
Depreciation	25,944	0.4%	25,944	0.4%	25,944							
Rent	39,452	0.6%	41,424	0.6%	42,667							
Health Insurance	102,202	1.5%	107,312	1.5%	115,360							
Shop Supplies	20,479	0.3%	22,381	0.5%	24,060							
Payroll Taxes	121,715	1.8%	127,192	1.7%	132,662							
Total Plant Exp	360,046	5.4%	379,533	5.2%	401,500	5.1%						
Total Cost of Sales	5,857,810	87.7%	6,217,482	85.2%	6,601,025	84.1%						
Gross Profit	821,899	12.3%	1,082,518	14.8%	1,246,475	15.9%	1,339,960	15.9%	1,440,457	15.9%	1,548.,492	15.9%
General & Administrative Expenses												
Salaries (Incl. Plt. Mgr)	168,330	2.5%	198,420	2.7%	250,000		250,000		250,000		250,000	
Payroll Taxes	18,862	0.3%	22,223	0.3%	22,223		22,223		22,223		22,223	
Commissions	80,444	1.2%	86,983	1.2%	93,485		100,497		108,034		116,136	
Advertising & Promotion	25,611	0.4%	26,000	0.4%	30,000		30,000		30,000		30,000	
Auto Expenses	16,819	0.3%	18,000	0.2%	18,000		18,000		18,000		18,000	
Misc Txs & Licenses	4,235	0.1%	3,500	0.0%	3,500		3,500		3,500		3,500	
Office Expense / Postage	22,978	0.3%	24,000	0.3%	25,800		27,735		29,815		32,051	
General Insurance	0	0.0%	0	0.0%	0		0		0		0	
Repairs & Maintenance	976	0.0%	1,200	0.0%	1,200		1,200		1,200		1,200	
Dues & Subscriptions	2,280	0.0%	2,400	0.0%	2,400		2,400		2,400		2,400	
Travel, Meals & Entertainment	35,843	0.5%	36,000	0.5%	36,000		36,000		35,000		36,000	
Telephone & Utilities	15,500	0.2%	16,021	0.2%	17,223		18,514		19,903		21,396	
Miscellaneous	41,000	0.6%	2,400	0.0%	2,400		2,400		2,400		2,400	
Total Gen & Admin	432,878	6.5%	437,127	6.0%	502,231	6.4%	512,469	6.1%	523,475		535,307	
Income from Operations	389,021	5.8%	645,390	a:s%	744,244		827,491		916,982		1,013,185	
Other Income & Deducts												
Interest Expense	0	0	0	0	0		0		0		0	
Bad Debts	56,000	0	36,500	0.5%	39,238		42,180		45,344		48,745	
Total Other Income / Deducts	(56,000)	(0)	(36,500)	-0.5%	(39,238)		(42,180)		(45,344)		(48,745)	
Pre- tax Income	333,021	5.0%	608,890	8.3%	705,006		785,311		871,638		964,441	
Income Taxes 35%	116,557	1.7%	213,112	2.9%	246,752		274,859		305,073		337,554	
Net Income	216,464	3.2%	395,779	SA%	458,254		510,452		566,565		626,886	
Add: Depreciation			25,944		25,944		25,944		25,944		25,944	
Less: net Add'l W/C	$683,440		51,258		51,258		51,253		51,258		51,258	
Net Cash Net income	Balance		370,465		432,940		485,138		541,251		601,572	

OCF Calculation; 25% R.ate of Return to determine Net Income forcing the Investment (purchase price)

	Investment	y1		y2	y3	y4	y5		y6	y7	y8	y9	y10
Cash Net Income	(1,750,000)	370,465		432,940	485,138	541,251	601,572		601,572	601,572	601,572	601,572	601,572

DCF   25.4%

In the above chart, we have taken the adjusted actual earnings and the projection (accepted by the company buying the business) and further projected it to a full five-year forecast. The projections are made by the various disciplines associated within the acquiring company. Thus, the Sales Department forecasts its part, as does the Production Operation and Purchasing Departments. Expenses are also forecast, on a line-by-line basis, by the managers involved. This is from the viewpoint of the purchasing company. However, a seller may want to put himself in the place of the purchaser to also value his company "on their basis," prior to negotiations. Thus, this exercise is potentially for both parties.

In this example, the annual growth rate of sales was determined to be 7.5% per year, above the accepted Year 1 forecast. Certain material and direct labor production efficiencies were included. Expenses generally remained stable except that in Year 2, additional salaried personal were added. Bad debts remained at .5% of sales. As a result Net Income was determined. To determine the annual net cash flows the business would generate, depreciation was "added back" to the operating Net Income, and additional working capital necessary to operate the growing business was subtracted. The Balance Sheet in Chapter Ten detailed an actual closing balance of accounts receivable of $890,880, inventory of $548,080, and accounts payable of $755,520; net $683,440. The incremental growth of working capital was assumed to match the rate of sales growth. Thus, additional cash would have to be put into the business to support the sales (a reduction to cash flow). Note, the investment also included the "buyout" of the working capital, which is considered as a substitution of cash for assets with no impact on the investment. Will the operating cash flows provide a sufficient return on the purchase price (investment)?

Note: Any change to the purchase price can adjust the above and impact the rate of return. For example, the seller may want to retain the accounts receivable. The buyer may counter that the seller must also retain the accounts payable (generally, the balances are not equal, with A/R being in excess of A/P.) Again, anything that isn't illegal is legal and can be negotiated.

The projection incorporates holding the fifth year Cash Net Income for Years 6 through 10 as most companies use such a time frame. Thus, investing $1,750,000 to produce an annual cash flow of $370,465, $432,940, etc., will result in a DCF rate of return of 25.4% given the assumptions incorporated. "What happens if" analyses usually follow in order to view several sets of assumptions: Sales volume; selling price (via increasing the COGS ratio); production efficiencies and other items. In addition, if only a five-year time frame were used, the purchase price to generate a 25% DCF would be $1,250,000; if a 20% yield were required over a five-year period, the investment could be as much as $1,400,000. Other scenarios can be tested, including testing the yields on the Net Earnings only (before testing the yields on depreciation and additional working capital items). However, the $1.7 million investment requirement is the most likely scenario for a healthy 25% required return.

Summary

In general, both methods presented support and are the basis for the method. The Earnings Multiple Method appears to have been derived (or proven) by the DCF method. The "unsophisticated" multiple and "detail" DCF methods are the most used by the unsophisticated and the sophisticated investor. Again, the goal is to maximize wealth.

CHAPTER NINETEEN —
MANAGEMENT CONTROLS

There are various areas of management control. We will address mostly financial controls in this last chapter. The key to any control system is checks and balances, and separation of duties within and outside of any discipline! However, that is not always possible, practical, or cost-effective.

Cash

The first and most prominent management control is CASH. Cash runs the company and it is the first target of theft. There are several activities that provide some means of assurance that it is sound and not being stolen. The first is the bank statement. **Depending upon the size of the company, the bank statement must be sent to, opened, and reviewed in detail by the owner.**

There are many small companies where there is an office assistant who probably calculates the payroll from time cards and disburses it, deposits the collections, invoices the customers may actually place the orders, ensures that what is ordered is received, ensures that the invoice from the vendor is correct, cuts the check, replenishes petty cash, reimburses expenses, is the receptionist, and performs many other functions.

How often do we read in the newspapers that someone in this type of position has been caught stealing from the company? "Oh, what a surprise, the assistant was such a trusted, faithful and hardworking employee. We just can't believe it." Doesn't this sound familiar?

There are numerous ways for theft to occur. As stated above, the first means to discover or prevent it is to have the bank statement sent to the owner, personally. The bank statement must be presented unopened (ensuring that all checks sent by the bank are there) when you receive it. Every check must be examined again, examine every check! Look for a forged signature, too many expense reimbursements, more than a reasonable number of disbursements to one vendor, and in various amounts. Are there new vendors appearing? Who has the ability to set up new vendors in the computer system? Spot check the vendor disbursements and charges. Are all the checks there that are clearing the bank? Do the dates on the checks exactly match what appears in the vendor file? You may want to go back a month or two to see if there have been any "adjustments."

In the normal course of business, the bank statement must be reconciled. This will be made to reconcile the ending or beginning balance of each month. The reconciliation should encompass two calculations:

The first is the most complex:

From the bank statement:

1. Check their balance as of the close of the month.

2. Subtract the checks written, but not yet cleared (no matter when the checks were written, as long as they are considered to be eligible for payment).

3. Add the deposits made by the month-end, but are not on the statement (this is usually a timing difference).

4. Adjust the ledger or checkbook balance for all fees and charges not entered during the month.

My suggestion is that whenever a charge or credit (including monies the bank receives and deposits into your account from credit card transactions) is known during the month, other than normal deposits, and disbursements that item is then entered into the books. It just reduces the chance of something being forgotten and it does provide a more accurate running balance.

Voila! You are in sync with the bank! OH, you're not? Now the fun begins; searching down the differences in detail. It happens in all businesses, almost every month!

The differences may be due to items entered into the cash account by error, timing differences, bank error, and just a slew of other reasons. However, in the end, the book balance and the adjusted bank statement must match to the penny.

The **second reconciliation** is to add to the last checkbook month-end balance (which should also be the general ledger balance) all the deposits and subtract all the checks written, plus and minus all other charges and credits (e.g. interest). The balance must match the current month-end balance.

When the reconciliation is complete, a review should be made by someone other than the person making the reconciliation.

The Financial Plan

Another control device is the financial plan and subsequent comparative financial statements. This will reveal differences in the expense versus budget, line by line. It will also outline the sales differences and production differences versus the estimate (direct labor, direct materials, and production expenses). The gross profit margin differences could be caused by differences in product assortment versus forecast (if there are differences in margins by product), reduced selling prices, and again, production inefficiencies versus estimate. A one-month snapshot may not present a proper picture; a trend is a better means of determining a problem. However, each financial statement must be examined.

The interim daily or weekly reports will provide current information to better manage your company and suggest areas to question. Financial statements are very important. However, at best they are produced five to ten workdays after the end of the month.

Inventory

Annual, semi-annual, quarterly, or other periodic physical inventories must be taken. These should match the inventory reports and the books with any differences analyzed and entered into the books. In addition, periodic cycle counts should be taken to ensure the counts are correct. Shortages could mean efficiencies are not what they are expected to be, damaged goods are going unreported, miscounts are occurring because items are being stored in the wrong place, theft, or the wrong items are being used or reported as used in the operation. For example, if a red widget has to be used in a particular operation, but there is a quirk in the system and a clerk cannot get it through, but the system easily accepts a green widget as used, then the red is actually used, but the green is reported. The company may run out of red and never know why. The use of inventory tickets for the relief of inventory (signed) should confirm the day's usage in the various production reports, plus or minus what is left on hand.

Differences may also indicate that we sent out too much or were shorted on an order and receiving never noticed or looked the other way. Did the clerk accept a short order and receive a kickback from the vendor? And, there are more scenarios. For instance, if there are overages, is the customer being shorted?

In theory, the ending inventory should be as follows: opening inventory, plus receipts, less shipments (and other items detailed) equals closing inventory.

This formula is used by many companies to cost out their sales, i.e. opening inventory, plus receipts, less closing inventory equals the cost of goods sold. This formula should generate the cost of sales on the income statement, which again will be compared to the financial plan.

Locks should be employed where necessary. They may be employed in general use and for special high value or critical items. And, let us not forget an alarm system for the property, which will be required by your insurance carrier.

Accounts Receivable

Reconciliation of the A/R balances must be made monthly. This will ensure that all monies received are applied properly. The person making the deposits should not be the individual applying the cash nor receiving or making the calls for past due invoices. This will ensure no mishandling or diverting of funds. A review of the Aging Report (which is a listing of each invoice by the customer and by invoice date, separated into columns usually in 30-day increments) is also necessary. Decide who makes the review. This graphically presents a spread of the Days Outstanding (previously discussed). This will prompt the initial calls.

Accounts Payable

This has a similar Aging Report and will prompt questions regarding why invoices were paid late or early. Is the company in a position to disburse early and take any discounts offered? Again, a month-end reconciliation is in order:

1. Detail sub-ledger to general ledger balance, and

2. Opening balance plus new invoices less trade disbursements equals new balance.

Also, for each disbursement **a three-way match** must be made: purchase order and receiving document, matched to the invoice. (Is the person ordering the items, the same who is receiving it and paying for it?) Then the invoice is entered into the books for payment. Thus, a correct liability is reported and the aging and due date is set up for payment, the opposite of the A/R operation.

Customer Distribution

Depending upon the business, it may not be wise to have too big a percent of the company's business with one customer. Sometimes it is unavoidable. However, a good spread of business is a good goal. I was the controller of a company that had its largest customer fail with our company having a $50,000 receivable balance. It was a tough loss. However, the customer represented less than 3% of our total business, so we were safe. Had this customer been 25% of our business, the company would have been in real trouble. I hope you get the picture.

In summary, as you go through your company's physical plant, general operations and administration, one must question what information might be helpful in managing the company and in preventing fraud. However, this must be reviewed with separation of duties in mind, the delegation of authority, and the time and cost necessary to prevent fraud and manage the company.

SUMMARY

The goal of this book was to educate the reader on the basics of finance for running a business. To actually execute this function, it may be necessary to employ a finance professional. This professional is usually your accountant. However, while all accountants should have been schooled in the techniques discussed within this book, they may not have sufficient experience in performing the functions described here. GIVE HIM OR HER THE BOOK!!!! EVEN BETTER, BUY ANOTHER COPY.

The education received from this book will enable the reader to ask:

1. Is the particular function being performed at your company?

2. Why should it or shouldn't it be performed?

3. What are the details of the function?

Then, you will understand the following:

1. What that function is supposed to do and how it is supposed to operate.
2. The background of the information given to you.
3. The result and meaning of it.
4. Through an understanding of each function you will be able to make better business decisions based upon that information, and it will enable you to better manage and control your company.

This book commenced with the basics of double-entry bookkeeping. This was input to inform the reader as to what actually happens in the accounting process; thereby, dissipating the cloud as to how it actually works. It then proceeded to define and group accounts in an organized manner for understanding, reporting, and to meet the legalities of operating a business. These groupings then resulted in creating financial statements to report how the company is doing and what it is worth. Depreciation (a NON-cash expense) was discussed and clarified, as well as the several legal structures under which businesses generally operate.

With this as a foundation, Section Two moved the reader to a higher level, interpreting the basics for better management to:

- Forecast the future
- Analyze results
- Compare performance

This led to one of the most critical subjects, detailing the cash flow of the company. In this Section, we discussed Working Capital, what it is, and the advantages and disadvantages of certain financial and operating policies under which a company may choose to operate. Also discussed were financial measuring tools that further the understanding of what your reports are revealing and enabling better company management. The Section then addressed review, forecasting, and comparison activities further enhancing one's ability to get one's arms around the company. When management gets involved in this process (planning and forecasting) as to why certain things happen, the business becomes even clearer. And then there is cash flow. This is the center of the universe. No Cash, no business! This chapter detailed the mechanics of what happened (positive and negative) with the company's money, and what is expected to happen. Thus, it places management in a position to plan for the needs or to be prepared to take advantage of an excess cash position.

As the reader then had an understanding regarding better business, and financial management, and planning, we then provided certain analytical tools including:

- Break-even and growth potential
- The creation and analysis of product costs
- Whether a business investment would be worthwhile
- Whether to lease or buy an asset, including the ramifications of each

- Issuing credit to a customer
- Management controls

Finally, what is the true potential market value of the company? (It is assumed that the company's ownership/stock is not regularly traded. If it were, then, this would be how the free market sets the company's value; via the price of the traded stock.)

One point that should be mentioned, should your company contemplate a new business strategy or investment and the analysis is positive, this "new piece" of the business will eventually materialize on your financial reports. Your company should create a longer-range strategic plan, annually covering all the disciplines elements of the current and future business. Reiterating, incremental analyses are made, and are "added to" the existing business to create a new complete picture, including all the cash impacts and total cash flow of the business. This may eventually require additional funding; debt or equity is determined by various company goals and ratios forecasted. It all comes into play!

Now that you read the book, enjoy your business.

Harvey Goldstein

GLOSSARY

Accounting	A system to manage and report all activities in terms of financial units; e. g. $ in the USA.
Accounts Payable	Items received and not yet paid
Accounts Receivable	Items billed to customers and not yet collected.
Accrual	An amount put into the accounting system representing value received, but without a bill from the vender.
Acid Test Ratio	A ratio representing assets which can be quickly converted to cash to be matched against those current liabilities
Amortization	A non-cash charge against earnings representing a portion of the value of a non-tangible asset.
Asset Turnover	The number of times sales are produced from the company's assets.
Balance Sheet	A financial statement reflecting the assets, liabilities and equity of a company at a point in time.
Bank Reconciliation	A system to match the amount on the books of the company to the amount stated on their bank statement.
Bankruptcy	Legal protection against creditors when a company cannot meet its obligations.
Break-even	The amount of sales needed to match costs and expenses resulting in no profit or loss.
Budgeting	A system of projecting sales, costs and expenses for a future period.
Business Investment	The expenditure of non operating funds to enhance the business.
Capital Budgeting	A system to determine which business investment projects to choose.
Cash	Funds in actual currency or in a demand deposit at a financial institution.
Cash flow	The net amount of funds coming into the business and leaving the business.
Cash Flow Statement	A financial statement reflecting where a company's funds were generated and where they were spent.
Cash Forecasting	A system to project the sources of business funds and their disbursement.

Chart of Accounts	The organization of the accounts in an accounting system.
Common Stock	A Security representing ownership in a company; having the lowest priority and highest opportunity for wealth improvement.
Controls	A system ensuring constant monitoring of an operation.
Corporation	A legal entity.
Cost of Capital	The price stated in terms of an interest rate, that a company must pay to secure the use of funds.
Cost Reduction	The ability to lower the cost to produce something.
Credit and Collections	The system of enabling customers to secure a company's goods or services and pay for them at a later date; and the company's method of securing payment.
Current Ratio	Current assets divided by current liabilities.
Days Sales Outstanding (DSO)	The average number of days to collect an account receivable.
DCF	See Discounted Cash Flow.
Debt Ratio	The ratio of all liabilities divided by all assets.
Depreciation Direct Labor	A non-cash charge against earnings representing a portion of the value of a tangible asset.
Direct Labor	The amount of money spent for human operations to actually produce a good or service.
Direct Materials	The amount of money spent for incoming items which will be combined to actually produce a good or service.
Discount Period	The number of days allowed to a customer wherein a reduced amount will be accepted for payment of an invoice.
Discounted Cash Flow (DCF)	A technique that utilizes the time value of money to measure the value of an investment project.
Discounts	An amount, usually expressed in percentage terms, that will reduce a payment.
Earnings Multiple	A factor used to multiply the defined earnings of a company to arrive at a company's market value.

Earnings Statement	A financial statement that financially describes the performance of a company over a period of time
Expansion	The growth of a company resulting from investment.
Financial Lease	A contract wherein a company has the use of an asset for a period of time, in excess of one year, under specific terms of use and payment.
Financial Planning	A system of projecting a company's total operation in the future, incorporating all incomes and expenditures.
Financial Ratios	A series of independent calculations, usually expressed in terms of percents, used to measure specific areas and total company efficiencies.
Financial Statements	The Balance Sheet, Income Statements and Cash Flow, Statements of a company
Fixed Accounts	Those accounts that will not change with a change in business volume.
Fixed Asset Turnover	An efficiency ratio that reveals the number of times sales are produced from the fixed assets.
Full Costs	The total costs associated with a product, including all direct costs, allocated share of production overhead and a share of general and administrative expense to cost the product down to the pre-taxearnings level.
General Partner	One of the owners in a legal partnership, having no special exemptions or privileges.
Gross Margin	The percent of sales generated by the profit after the cost of sales is subtracted from the sales.
Income Statement	See Earnings Statement
Internal Rate of Return (IRR)	An efficiency ratio that reveals the number of times inventory is shipped as the cost of sales.
Inventory Turnover	A procedure to determine the ramifications of an investment from a financial perspective.
Investment Analysis Lease	A contract that enables the use of an asset by an individual who doesn't own it; for a price.
Lease Vs. Purchase	An analysis to determine the least costly means to acquire an asset
Limited Partner	One of the owners in a legal partnership; however, this individual is not involved in the management of the business, thus, the liability of this partner is limited to the original investment.

LLC	Limited Liability Corporation; a legal form of corporation.
LLP	Limited Liability Partnership; a legal form partnership
Management Controls	A system wherein a business sets procedures in place to ensure information flow and provide the ability to detect problems.
Marginal Costs	The cash cost of a product without any allocation for indirect, fixed or other allocated expense.
Net Present Value	The value of future cash flows brought to the current value by discounting these flows by a specific interest rate, less the original outlay of funds.
Net Profit Margin	The percent generated by dividing the Net Profit (after income taxes) by the Sales. Net percent profitability.
Net Working Capital	Current assets less current liabilities; a positive amount available after all current assets are converted to cash, less the payment of current obligations.
Operating Lease	The use of an asset under a contract that can be cancelled upon short notice.
Operating Margin	The percent generated by dividing the operating earnings by the sales; the profitability from regular business operations.
Profit and Loss Statement	See Earnings Statement
Partnership	A legal company organization wherein all investors own a share of the business without the taxes or legal protections of a corporation. Almost a sole proprietorship with several owners.
Payback Period	The amount of time for an investment to generate the cash that matches the original investment.
Preferred Stock	An equity security of a corporation that has higher priority than common stock and a "guaranteed" return; but as such will not, generally, enjoy significant appreciation nor reduction in value.
Present Value	The value of future cash flows brought to the current value by discounting these flow by a specific interest rate.
Present Worth	See Present Value
Product Cost	The amount of money required to produce a unit of goods; usually including all materials, labor and production overhead.
Project Return	See Discounted Cash Flow.

Quick Ratio	The same as the Current Ratio with Inventory excluded from the assets. Thus, the ability to very quickly generate cash "to pay" the current liabilities.
Rate of Return	The amount received as a percent of the amount invested. This could be static as ROE below or dynamic as in DCF.
Reconciliation	A matching of one set of data against another set of data.
Return on Assets	The amount of money earned measured against the amount of assets used to generate the earnings.
Return on Equity (ROE)	The amount of money earned relative to the Net Worth of the business; the NW is the business' net value.
S Corporation	See Sub chapter S corporation
Seasonality	The sales pattern of a company throughout a year's period.
Sole Proprietorship	A company wherein the ownership is by one person; having no legal protections and earnings are only taxed once.
Standard Cost	The expected cost of a unit (see Product Cost) for the forthcoming period.
Sub Chapter S Corporation	A legal entity of 35 or less investors enjoying corporate legal protections, without having to pay taxes on its own.
Sustainable Growth Rate	The ability of a company to grow, utilizing internal funds only, expressed in terms of a percentage.
Times Interest Earned	The ability of a company to pay the debt interest from operating earnings; usually expressed as a ratio to "1": e. g. "X:1".
Time Value of Money	A concept wherein funds secured or paid in the future enjoy greater or lesser value over the period, at an interest rate.
Valuation of a Company	How much a company is worth in an open market. This can be estimated based upon several different bases.
Variable Accounts	The costs from certain accounts. These accounts costs will change with a change in business conditions.
Variations	Differences from expected values
Working Capital	The value of a company's current assets.....See Net Working Capital.

APPENDICES/TABLES

Present Worth of $1 received at the end of n number of years at x percent interest rate

year	1%	2%	3%	4%	5%	6%	7%	8%	9%	10%	11%	12%	13%	14%	15%	18%	20%	24%
1	0.990	0.980	0.971	0.962	0.952	0.943	0.935	0.926	0.917	0.909	0.901	0.893	0.385	0.377	0.870	0.847	0.833	0 806
2	0.880	0.961	0.843	0.925	0 907	0.890	0.873	0.857	0.842	0.826	0.812	0.797	0.783	0.769	0.756	0.718	0.694	0.650
3	0.971	0.942	0.915	0.889	0.864	0.540	0.816	0.794	0.772	0.751	0.731	0.712	0.693	0.675	0.658	0.609	0.579	0.524
4	0.961	0.924	0.888	0.855	0.823	0.792	0.763	0.735	0.708	0.683	0.659	0.636	0:613	0.592	0.572	0.516	0.482	0.423
5	0.951	0.906	0.863	0.822	0.784	0.747	0.713	0.681	0.650	0.621	0.593	0.567	0.543	0.519	0.497	0.437	0.402	0.341
6	0.942	0.888	0.837	0.710	0746	0.705	0.666	0.630	0.596	0.564	0.535	0.507	0.480	0.456	0.432	0.370	0.335	0.275
7	0.833	0.871	0.813	0.760	0.711	0.665	0.623	0.583	0.547	0.513	0.482	0.452	0.425	0.400	0.376	0.314	0.279	0.222
8	0.923	0.853	0.788	0.731	0.677	0.627	0.582	0.540	0.502	0.467	0.434	0.404	0.376	0.351	0.327	0.266	0.233	0.179
9	0.914	0.837	0.766	0.703	0.645	0.592	0.544	0.500	0.460	0.424	0.391	0.361	0.333	0.308	0.284	0.225	0.194	0.144
10	0.805	0.820	o:744	0.676	0.614	0.558	0.508	0.463	0.422	0.386	0.352	0.322	0.295	0.270	0.247	0.191	0.162	0.116
11	0.896	0.804	0.722	0.650	0.585	o:s21	0.475	0.429	0.388	0.350	0.317	0.287	0.261	0.237	0.215	0.162	0.135	0.094
12	0.887	0.788	0.701	0.625	0.557	0.497	0.444	0:397	0.356	0.319	0.286	0.257	0.231	0.208	0.187	0.137	0.112	0.076
13	0.879	0.773	0&81	0.601	0.530	0.469	0.415	0.368	0.326	0.290	0.258	0.229	0.204	0.182	0.163	0.116	0.093	0:061
14	0.870	0.758	0.661	0.577	0.505	0.442	0.388	0.340	0.299	0.263	0.232	0.205	0.181	0.160	0.141	0.099	0.078	0.048
15	0.861	0.743	0.642	0.555	0.481	0.417	0.362	0.315	0.275	0.239	0.209	0.183	0.160	0.140	0.123	0.084	0.015	0.040
16	0.853	0 728	0.623	0.534	0.458	0.394	0.339	0.292	0.252	0.218	0.188	0.163	0.141	0.123	0.107	0.071	0.054	0.032
17	0.844	0.714	0.605	0.513	0.436	0.371	0.317	0.270	0.231	0.198	0.170	0.146	0.125	0.108	0.093	0.060	0.045	0.026
18	0.836	0.700	0.587	0.494	0.416	0.350	0.296	0.250	0.212	0.180	0.153	0 130	0.111	0.095	0.051	0.051	0.038	0.021
19	0.828	0.686	0.570	0.475	0.396	0.331	0.277	0.232	0.194	0.164	0.138	0.116	0.098	0.083	0.070	0.043	0.031	0.017
20	0.820	0.673	0.554	0.456	0.377	0.312	0.258	0.215	0.178	0.149	0.124	0.104	0.087	0.073	0.061	0.037	0.026	0.014

The value of $1 investecfnow at the end of n number of years at x percent interest rate
This is $1 compounded each year at the interest rate.

year	1%	2%	3%	4%	5%	6%	7%	8%	9%	10%	11%	12%	13%	14%	15%	18%	20%	24%
1	1.010	1.020	1.030	1.040	1.050	1.060	1070	1.088	1.090	1.100	1.110	1.120	1.130	1.140	1.150	1.180	1.200	1.240
2	1.020	1.040	1,061	1.082	1.103	1.124	1.145	1.166	1.188	1.210	1.232	1.254	1.277	1.300	1.323	1.392	1.440	1.538
3	1.030	1.061	1.093	1.125	1.158	1.191	1.225	1.260	1.295	1.331	1.368	1.405	1.443	1.482	1.521	1.643	1.728	1.907
4	1.041	1.082	1.126	1.170	1.216	1.262	1.311	1.860	1.412	1.464	1.518	1.574	1.630	1.689	1.749	1.939	2.074	2.364
5	1.051	1.104	1.159	1.217	1.276	1.338	1.403	1.469	1.539	1.611	1.685	1.762	1.842	1.925	2.011	2.288	2.488	2.932
6	1.062	1.126	1.194	1.265	1.340	1.419	1.501	1.587	1.677	1.772	1.870	1.974	2.082	2.195	2.313	2.700	2.986	3.635
7	1.072	1.149	1.230	1.316	1.407	1.504	1.606	1.714	1.828	1.949	2.076	2.211	2.353	2.502	2.660	3.185	3.583	4.508
8	1.083	1.172	1.267	1.369	1.477	1.594	1.718	1.851	1.993	2.144	2.305	2.476	2.608	2.853	3.059	3.759	4.300	M90
9	1.094	1.195	1.305	1.423	1.551	1.689	1.838	1.999	2.172	2.358	2.558	2.773	3.004	3.252	3.518	4.435	5.160	6.931
10	1.105	1.219	1.344	1.480	1.629	1.791	1.967	2.159	2.367	2.594	2.839	3.106	3.395	3.707	4.046	5.234	6.192	8.594
11	1.116	1.243	1.384	1.539	1.710	1.898	2.105	2.332	2.580	2.853	3.152	3.479	3.836	4.226	4.652	6.176	7.430	10.657
12	1.127	1.268	1.426	1.601	1.796	2.012	2.252	2.518	2.813	3.138	3.498	3.896	4.335	4.818	5.350	7.288	8.916	13.215
13	1.138	1.294	1.969	1.665	1.886	2.133	2.410	2.720	3.066	3.452	3.883	4.363	4.898	5.492	6.113	9.599	10,699	16.386
14	1.149	1.319	1.513	1.732	1.980	2.261	2.679	2.987	3.342	3.797	4.310	4.887	5.535	6 261	7.076	10.147	12.839	20,319
15	1.161	1.346	1.558	1.801	2.079	2.397	2.759	3.172	3.642	4.177	4.785	5.474	6.254	7.138	8.137	11.974	15.407	25.196
16	1.173	1.373	1.605	1.873	2.183	2.540	2.952	3.426	3.970	4.595	5.311	6.130	7.067	8.137	9.358	14.129	18.488	31.243
11	1.184	1.400	1.653	1.948	2.292	2.693	3.159	3.700	4.328	5.054	5.895	6.866	7.986	9.276	10.761	16.672	22.186	38.741
18	1.196	1.426	1.702	2.026	2.407	2.854	3.380	3.996	4.717	5.560	6.544	7.690	9.024	10.575	12.375	19.673	26.623	48.039
19	1.208	1.457	1.154	2.107	2.527	3.026	3:617	4.316	5.142	6.116	7.263	8.613	10.197	12.056	14.232	23.214	31.948	59.568
20	1.220	1.486	1.806	2.191	2.653	3.207	3.870	4.661	5.604	6.727	8.062	9.646	11.528	13.743	16.361	27.893	38.338	73.864

This is the receipt ofthe same amount each year for a given number of years; an annuity
Present Worth of$1 received at the end of each year for n number of years at x percent interest rate

year	1%	2%	.J%2				7%			1.llJi	.l.1li	m:2	m	1!lli	illi	.1lffi	D	.2::!li
1	0.990	0.980	0.971	0.962	0.952	0.943	O.S35	0.926	0.917	0.909	0.901	0.893	0.885	0.877	0.870	0.847	0.833	0.806
2	1.970	1.942	1.913	.1.886	1.859	1.833	1.808	1.783	1.759	1.736	1.713	1.690	1.668	1.647	1.626	1566	1.528	.1457
3	2.941	2.884	2.829	2.775	2.723	2..673	2.624	2:577	2.531	2:487	2.444	2.402	2.361	2.322	2.283	2.174	2.106	1.981
4	3.902	3.808	3.717	3.630	3.546	3.465	3.387	3.312	3.240	3.170	3.102	3.037	2.974	2.914	2.855	2.690	2.589	2404
5	4.853	4.713	4.580	4452	4.329	4.212	4.100	3.993	3.890	3.791	3.696	3.605	3.517	3433	3.3½2	3.127	2.991	2.745
6	5.795	5.601	5:417	5.242	5.076	4.817	4.767	4.623	4:486	4.355	4.231	4.111	3.998	3.889	3.784	3:498	3.326	3.020
7	6.728	6.472	6230	6.002	5.786	5.582	5.389	5.206	5.033	4.868	4.712	4.564	4.423	4.288	4.160	3.812	3.605	3.242
8	7652	7.325	702.0	6.733	6463	.6.210	5.971	5.747	5.535	5.335	5.146	4.968	4.79[]	4.639	4.487	4.078	3.837	3421
8	8.566	8.162	7.786	7:435	7.108	6.802	6.515	6.247	5.995	5:759	5.537	5.328	5.132	4.946	4.772	4.303	4.031	3.566
10	9.471	8.983	8.530	8.111	7,722	7.360	7.024	6.710	6418	6.145	5.889	5.650	5426	5.216	5.019	4.494	4:192	3.682
11	10.368	g 787	9253	8.760	8.306	7.887	7.499	7 139	6805	6.495	6.207	5.838	5.687	5.453	5.234	4.656	4.327	3.776
12	11.255	10.575	9.954	9.385	8.863	8.384	7.843	7.536	7.161	6.814	6.492	6.194	5.918	5.660	5.421	4.793	4.439	3.851
13	12.134	11 348.	10.635	9986	9.394	8.853	8.358	7.904	7.487	7.103	6.750	6.424	6.122	5.842	5.583	4.910	4.533	3.912
14	13.004	12.106	11.296	10.563	9899	9.295	8.745	8.244	7.186	7.367	6.982	6.628	6.302	6.002	5.724	5.008	4.611	3.962
15	13.865	12.849	11.938	11.118	10.330	9.712	9.108	8.559	8.061	7.606	7.191	6:811	6.462	6.142	5.847	5.092	4.675	4.001
16	14.718	13.578	12.561	11.652	10.838	10.106	9.447	8.851	8.313	7.824	7.379	6.974	6.604	6.265	5.954	5.162	4.730	4.033
17	15.562	14.292	13166	12.166	11.274	10477	9.763	9.122	8.544	8.022	7.549	7.120	6.729	6373	6.047	5.222	4.775	4.059
18	16.398	i4.992	13.754	12.659	11.690	10.828	10.059	9.372	8.756	8:201	7.702	7.250	6.840	6.467	6.128	5.273	4.812	4.080
19	17.226	15.678	14:324	13.134	12085	11.158	10.336	9.604	8.950	8.365	7,839	7366	6.938	6.550	6.198	5.316	4.843	4.097
20	18.046	16.351	14.877	13.590	12462	11A70	10.594	9818	9.129	8.514	7.963	7.469	7.025	6.623	6.259	5.353	4.870	4.110

The Present Worth of$1 invested each year for n years at x interest rate
The Present Worth ofanAnnulty

year	1%	2%	3.%	4%	5%	6%	7%	8%	8%	10%	11%	12%	13%	14%	15%	18%	20%	24%
1	1.000	1.000	1.000	1.000	1.000	1.000	1.000	1.000	1.000	1.000	1.000	1.000	1.000	1.000	1.000	1.000	1.000	1.000
2	2.010	2.020	2.030	2.040	2 050	2.060	2.070	2.080	2080	2.100	2.110	2.120	2:130	2.140	2.150	2.180	2.200	2.240
3	3.030	3.060	3.091	3.122	3.153	3.184	3.215	3.246	3.211!	3.310	3.342	3.374	3.407	3.440	3.473	3.572	3.640	3.778
4	4.060	4.122	4.184	4.246	4.310	4.375	4.440	4.506	4.573	4.641	4.710	4.778	4.850	4.921	4.893	5.215	5.368	5.684
5.	5.101	5.204	5.308	5.416	5.526	5.687	5.751	5.867	5.985	6.105	6.228	6.353	6.480	6.610	6.742	7.154	7.442	8.048
6	6.152	6.308	6.468	6.633	6.802	6.975	7.153	7.336	7.523	7.716	7.913	8.115	8.323	8.536.	8.754	9.442	9.930	10.980
7	7.214	7.434	7.662	7.898	8.142	8.394	8.654	8.923	9.200	9.487	8.783	10.089	10.405	10.780	11 087	12:142	12.916	14.615
8	8.286	8.583	8.892	9.214	8.549	9.897	10.260	10.637	11.028	11.436	11.859	12.300	12757	13.233	13.727	15.327	16.499	19.123
9	8.369	9.755	10.158	10.583	11.027	11.491	11.978	12.488	13.021	13.579	14.164	14.776	15.416	16.085	16.786	18.036	20.798	24.712
10	10.462	10.950	11.464	12.006	12.578	13.181	13.816	14.487	15.193	15.937	16.722	17.549	18.420	19.337	20.304	23.521	25.959	31.643
I1	11.567	12.168	12.808	13.486	14.207	14.972	15.784	16.645	17.560	18.531	19.561	20.655	21.814	23.045	24.349	28.755	32.150	40.238
12	12.683	13.412	14.192	15.026	15.817	16.870	17.888	18.977	20.141	21.384	22.713	24.133	25.650	27.271	28.002	34.831	39.581	50.895
18	13.809	14.680	15.681	16.627	17.713	18.882	20.141	21.485	22.953	24.523	26.212	28.029	29.985	32.089	34.852	42.219	48.997	64.110
14	14.947	15.974	17.086	18:292	19.598	21 015	22.550	24.215	26.019	27.975	30.095	32.393	34.883	37.581	40.505	50.818	59.196	80.496
15	16.087	17.293	18.599	20.024	21.579	23.276	25.129	27.152	29.361	31.772	34.405	37.280	40.417	43.842	47.580	60.965	72.035	100.815
16	17.258	18.639	20.157	21.825	23.657	25.673	27.888	30.324	33.003	35.950	38.190	42.753	46.672	50.980	55.717	72.939	87.442	126.011
17	18.430	20.012	21.762	23.698	25.840	28.213	30.840	33.750	36.974	40.545	44.501	48.884	53.739	59.118	65.075	87.068	105.831	157.253
18	19.615	21.412	23.414	25.645	28.132	30.906	38.898	37.450	41.301	45.588	50.386	55.750	61.725	68.394	75.836	103.740	128.117	195.994
19	20.811	22.841	25.117	27.671	30.539	33.760	37.378	41.446	46.018	51.158	56.939	63.440	70.749	78.868	88.212	123.414	154.740	244.033
20	22.019	24.297	28.870	29.778	33.066	36.786	40.995	45.762	51.160	57.275	64.203	72.052	80.947	91.025	102.444	146.628	186.688	303.601

ABOUT THE AUTHOR

I have been a chief financial officer or controller of several companies ranging from $5 million to $80 million. My responsibilities included all the accounting, analysis, reporting (to management, ownership and lending institutions), credit and collections, cash management and forecasting, financial analysis and planning, and strategic plan coordination. I was also a senior financial analyst with a NYSE listed company and a management consultant with two top consulting companies. I am, currently, the Chief Financial Officer of a ".com" company. I hold a BA in Economics from the City College of New York, City University of New York and an MBA (Finance) from The Bernard M. Baruch College, City University of New York. I am also an instructor in the Undergraduate and Graduate Schools of Business at the University of Phoenix.

During my career, I have arranged for loans, have created numerous financial plans and cash management systems. I have also performed financial analyses for major investment projects of up to $50 million, created product costs and profitability measures, and have incorporated all this into financial procedures manuals.

For recreation, I play at golf and I do some bicycle riding (note: neither Tiger Woods nor Lance Armstrong need to worry about me!)